Reducing Air Emissions Associated With Goods Movement: Working Towards Environmental Justice

NOVEMBER 2009

A Report of Advice and Recommendations
of the
National Environmental Justice Advisory Council
A Federal Advisory Committee to the U.S. Environmental Protection Agency

ACKNOWLEDGEMENTS

The National Environmental Justice Advisory Council (NEJAC) acknowledges the efforts of the Goods Movement Work Group (GMWG) in preparing the initial draft of this report. The NEJAC also acknowledges the stakeholders and community members who participated in the GMWG's study by providing public comments. Environmental justice communities, regulatory organizations, environmental groups, and other interested parties worked long and hard on this study. The staff of EPA's Office of Environmental Justice, especially Victoria Robinson, the GMWG's Designated Federal Officer, spent many hours meeting with the GMWG, ably assisted by EPA staff and ICF International, Inc, which provided contractor support.

DISCLAIMER

This Report and recommendations have been written as part of the activities of the National Environmental Justice Advisory Council, a public advisory committee providing independent advice and recommendations on the issue of environmental justice to the Administrator and other officials of the United States Environmental Protection Agency (EPA or the Agency). In addition, the materials, opinions, findings, recommendations, and conclusions expressed herein, and in any study or other source referenced herein, should not be construed as adopted or endorsed by any organization with which any Work Group member is affiliated.

This report has not been reviewed for approval by EPA, and hence, its contents and recommendations do not necessarily represent the views and the policies of the Agency, nor of other agencies in the Executive Branch of the Federal government.

NATIONAL ENVIRONMENTAL JUSTICE ADVISORY COUNCIL

Executive Council
- Richard Moore, Southwest Network for Environmental and Economic Justice (NEJAC Chair)
- Don Aragon, Wind River Environmental Quality Commission
- Chuck Barlow, Entergy Corporation
- Sue Briggum, Waste Management, Inc.
- Peter Captain, Sr., Tanana Chiefs Council
- Jolene Catron, Wind River Alliance
- Wynecta Fisher, New Orleans Mayor's Office of Environmental Affairs
- Jodena Henneke, Texas General Land Office
- M. Kathryn Brown, University of Cincinnati College of Medicine
- William Harper, Pacific Gas and Electric Company
- Christian Holmes, Global Environment and Technology Foundation
- Hilton Kelly, Community In-Power Development Association, Inc.
- J. Langdon Marsh, Portland State University, National Policy Consensus Center
- Gregory Melanson, formerly with Bank of America
- Paul Mohai, University of Michigan School of Natural Resources & Environment
- Patricia E. Salkin, Albany Law School
- Shankar Prasad, Coalition for Clean Air
- John Ridgway, Washington State Department of Ecology
- John Rosenthall, National Small Town Alliance
- Omega Wilson, West End Revitalization Association
- Elizabeth Yeampierre, UPROSE
- Victoria Robinson, NEJAC Designated Federal Officer (DFO), U.S. Environmental Protection Agency, Office of Environmental Justice

Goods Movement Work Group *
- Shankar Prasad, M.B.B.S, Coalition for Clean Air (*Work Group Co-Chair*)
- Terry Goff, Caterpillar Inc. (*Work Group Co-Chair*)
- Margaret Gordon, West Oakland Environmental Indicators Project
- Andrea Hricko, Community Outreach and Education Program, Southern California Environmental Health Sciences Center and Children's Environmental Health Center based at Keck School of Medicine, University of Southern California
- Angelo Logan, East Yard Communities for Environmental Justice
- Kirk Marckwald, California Environmental Associates **
- J. Langdon Marsh, National Policy Consensus Center, Portland State University
- Cynthia Marvin, California Air Resources Board
- Gregory Melanson, Formerly with Bank of America
- Wayne Grotheer, Port of Seattle
- Omega Wilson, West End Revitalization Association
- Victoria Robinson

The following individuals participated briefly in early discussions of the Work Group but were not part of the group that finalized the Report: Joyce King, Mohawk Council of Akwesasne; and Aston Hinds, Port of Houston.

*** Mr. Kirk Marckwald, California Environmental Associates, also served as a member of the GMWG from June 2007 through May 2009. His knowledge and experience gained from working with different railroads as a consultant to the American Association of Railroads (AAR) allowed Mr. Marckwald to actively participate in the meetings and deliberations of the GMWG. After the conclusion of the deliberations on this report and its recommendations, Mr. Marckwald informed the GMWG that his client, AAR, requested that Mr. Marckwald's name be deleted from the list of members of the GMWG. In addition, although there was no request for endorsement of the report by any organization with which any work group member is affiliated, the AAR informed the GMWG that AAR could not achieve consensus to endorse the report.*

NATIONAL ENVIRONMENTAL JUSTICE ADVISORY COUNCIL

November 20, 2009

Members:
Richard Moore, Chair
Don Aragon
Sue Briggum
Chuck Barlow
M. Kathryn Brown
Peter Captain, Sr.
Jolene Catron
Wynecta Fisher
William Harper
Jodena Henneke
Christian Holmes
Hilton Kelley
Langdon Marsh
Greg Melanson
Paul Mohai
Shankar Prasad
John Ridgway
John Rosenthal
Patricia Salkin
Omega Wilson
Elizabeth Yeampierre

Lisa P. Jackson
Administrator
U.S. Environmental Protection Agency
1200 Pennsylvania Avenue, NW (MC1101A)
Washington, D.C. 20460

Dear Administrator Jackson:

The National Environmental Justice Advisory Council (NEJAC) is pleased to submit the report, *Reducing Air Emissions Associated with Goods Movement: Working towards Environmental Justice* (November 2009), for the Agency's review. This report contains advice and recommendations about how the Agency can most effectively promote strategies, in partnership with federal, state, tribal, and local government agencies, and other stakeholders, to identify, mitigate, and/or prevent the disproportionate burden on communities of air pollution resulting from goods movement. With these recommendations, the Council wishes to:

- Provide a clear focus on the need to protect human health within communities impacted by exposure to air emissions from goods movement facilities and activities; and
- Convey a sense of urgency toward taking action for reducing exposure to air emissions in communities prioritized for action; and
- Emphasize differential approaches needed when addressing impact mitigation between existing and new goods movement facilities or activities.

The following is the list of key recommendations proposed by the NEJAC:

- Increase impacted communities' capacity and effectiveness to engage in and influence decisions related to goods movement that impact them;
- Direct each of the ten regions of EPA to identify and prioritize areas or communities maximally exposed or affected by goods movement related facilities and activities for taking action;
- Initiate mechanisms, processes and venues for reaching agreements on actions needed to reduce health impacts from goods movement in the identified communities;
- Accelerate introduction of existing, cleaner technologies and systems by providing needed resources using incentives, regulatory actions, modifying

A Federal Advisory Committee to the U.S. Environmental Protection Agency

existing funding and financing programs, creating new funding mechanisms, and offering technical assistance; and
- Support additional research and data gathering, with full community involvement and participation to accelerate emission reduction from goods movement activities.

Sincerely,

Elizabeth Yeampierre
Acting Co-Chair

John Ridgway
Acting Co-Chair

cc: Richard Moore, NEJAC Chair
NEJAC Members
NEJAC Goods Movement Work Group Members
Charles Lee, Director, Office of Environmental Justice (OEJ)
Victoria Robinson, NEJAC DFO, OEJ]

TABLE OF CONTENTS

1. **INTRODUCTION** 1
   Premise for Recommendations 1
2. **BACKGROUND** 3
   2.1 Scope of Goods Movement 3
   2.2 Air Pollution from Goods Movement 3
   2.3 Health Impacts Due to Air Pollution from Goods Movement 4
   2.4 Community Impacts and Environmental Justice 6
   2.5 Legal and Regulatory Environment 7
   2.6 Land Use Planning and Zoning 11
3. **FINDINGS AND RECOMMENDATIONS** 12
   3.1 Effective Community Engagement 12
      3.1a. Community Facilitated Strategies 13
      3.1b Collaborative Governance And Problem-Solving Strategies 15
   3.2 Health Research Data Gaps And Education Needs 18
   3.3 Regulatory And Enforcement Mechanisms 21
   3.4 Land Use Planning And Environmental Review 23
   3.5 Technology 25
   3.6 Environmental Performance, Planning, And Management 26
   3.7. Resources, Incentives, And Financing 28

APPENDICES

APPPENDIX A: Charge for Developing Recommendations to Address the Air Quality Impacts of Goods Movements on Communities

Appendix B: Glossary of Acronyms And Terms

Appendix C: Summary of EPA FACA Recommendations Regarding Air Quality From Freight Movement and Environmental Justice

A Federal Advisory Committee to the U.S. Environmental Protection Agency

THIS PAGE INTENTIONALLY LEFT BLANK

A Federal Advisory Committee to the U.S. Environmental Protection Agency

Reducing Air Emissions Associated With Goods Movement: Working Towards Environmental Justice

A Report of Advice and Recommendations of the National Environmental Justice Advisory Council

1. INTRODUCTION

The National Environmental Justice Advisory Council (NEJAC) is a Federal advisory committee chartered pursuant to the Federal Advisory Committee Act (FACA) to provide advice and recommendations to the Administrator of the U.S. Environmental Protection Agency (EPA or the Agency) about matters of environmental justice.[1] Environmental justice is the fair treatment and meaningful involvement of all people regardless of race, color, national origin, or income with respect to the development, implementation, and enforcement of environmental laws, regulations, and policies.

In June 2007, EPA requested that the NEJAC "provide advice and recommendations about how the Agency can most effectively promote strategies, in partnership with federal, state, tribal, and local government agencies, and other stakeholders, to identify, mitigate, and/or prevent the disproportionate burden on communities of air pollution resulting from goods movement."

The NEJAC asked EPA to establish the Goods Movement Work Group (GMWG) to research and identify potential recommendations. The GMWG was comprised of public and private sector stakeholders. A list of members is shown behind the title page of this report. The GMWG members met in person and via conference call on a regular basis to develop recommendations. The NEJAC also obtained public comments from additional stakeholders, including community groups, in public meetings on September 18, 2007, June 20, 2008, and October 21, 2008.

In accordance with EPA's request, the recommendations in this report primarily focus on methods to reduce air pollution from goods movement and its impacts on environmental justice communities near air and marine ports, rail yards, highways, bridges, border crossings, and distribution centers. The NEJAC understands that other environmental and quality-of-life issues exist due to goods movement activities. While such concerns are not directly addressed, some quality-of-life issues might be indirectly improved through the implementation of these recommendations (e.g., less truck idling at freight facilities will reduce both air pollution and noise).

Premise for Recommendations

Based on EPA's "Draft Charge for Developing Recommendations to Address the Air Quality Impacts of Goods Movement on Communities" (June 4, 2007), the NEJAC (and its Work Group) considered the following as the starting point to resolve controversies and formulate recommendations for further EPA actions:

There are serious public health concerns associated with goods movement due to high levels of air pollution and its associated health effects. The distribution of freight (goods movement) in the U.S. involves an entire system of transportation facilities, including seaports, airports, railways, truck lanes, logistics centers, and border

> **What is Goods Movement?**
>
> Goods movement refers to the distribution of freight (including raw materials, parts, and finished consumer products) by all modes of transportation, including marine, air, rail, and truck. Goods movement facilities, also called freight facilities, include seaports, airports, and land ports of entry (border crossings), rail yards and rail lines, highways and high truck traffic roads, and warehouse and distribution centers. The terms goods movement and freight transport are used interchangeably in this report.

---

[1] See www.epa.gov/environmentaljustice/nejac/

crossings. The vehicles and equipment that move goods today are predominantly powered by large diesel engines that emit particulate matter (PM), nitrogen oxides (NOx) that form ozone and fine particles in the atmosphere, hydrocarbons, and other air toxics. These air pollutants contribute to respiratory illness, heart disease, cancer, and premature death.

The environmental, public health and quality-of-life impacts of goods movement on communities are more pronounced in areas with major transportation hubs and high traffic roads. Minority and low-income communities near these hubs and throughways bear disproportionate impacts because of their close proximity to multiple pollution sources.

EPA asked the NEJAC to identify and summarize the most significant community environmental and/or public health concerns related to air pollution from goods movement activities. The Agency also suggested that the report address the types of data and tools that can be used to determine the location and magnitude of disproportionate impacts of air pollution related to goods movement activities on communities.

EPA has already made substantial efforts to reduce emissions from diesel engines, including those used for goods movement. These efforts include engine emission standards, incentives and other financial models, port emission inventories, and use of facility Environmental Management Systems.

Other government agencies, the freight industry, and affected communities have made progress in reducing diesel emissions from goods movement in many locations, but more must be done to meet health goals and fulfill EPA's commitment to ensure environmental justice. EPA suggests that the NEJAC "Specifically, identify the venues and other mechanisms that EPA can use to work with other government agencies, industry, and communities, in areas such as environment, public health, transportation, and/or land use, to reduce community exposure to air pollution from goods movement activities."

With this suggestion, EPA explicitly encouraged the NEJAC to expand the scope of recommended strategies to include not only what EPA can do under its own authority and funding, but also what EPA can accomplish by: (a) influencing other agencies at all levels of government; (b) leveraging change in the freight industry, and (c) empowering effective community involvement and action.

The charge implicitly recognizes the need to complement national (and international) actions with local and regional-scale strategies to further cut exposure in impacted communities. For example, land use and transportation infrastructure decisions can play a critical role in mitigating (or exacerbating) exposure to goods movement pollution in nearby communities– these decisions must be addressed in the recommendations.

This report is organized into three primary sections. Following this introduction in Section 1, Section 2 provides brief background about goods movement, air quality, health impacts, and the existing regulatory environment. Section 3 presents recommendations for each of seven focus areas where EPA can play a role in reducing goods movement air emission impacts. These focus areas include:

- Effective Community Engagement
    a. Community facilitated strategies
    b. Collaborative Governance
- Health Research Gaps and Educational Needs
- Regulatory and Enforcement Mechanisms
- Land Use
- Technology
- Environmental Management and Performance
- Financing

Appendix A provides the EPA charge to the NEJAC as well as NEJAC's charge to the Goods Movement Work Group. Appendix B provides a list of acronyms and a glossary of key terms. Appendix C includes a list of related recommendations, prepared by other EPA federal advisory committees, which relate to environmental justice and air quality from freight movement.

2. BACKGROUND

This section provides an overview of the current and future scope of goods movement operations in the United States, as well as the resulting air pollution emissions and regulatory structures to address those emissions.

2.1 Scope of Goods Movement

The U.S. has an extensive network of infrastructure to support goods movement, including highways, bridges, border crossings, air and marine ports, rail lines, rail yards, and distribution centers. Goods movement activities have increased significantly in the past 20 years. Container shipments quintupled at the ten largest U.S. container ports from 1980 to 2006, and over the last decade alone, shipments have grown by 81 percent.[2] The Federal Highway Administration (FHWA) forecasts that between 2006 and 2035:

- Freight tonnage hauled by trucks will grow by 80 percent;
- Rail tonnage hauled will grow by 73 percent;
- Water transportation tonnage will increase by 51 percent;
- Intermodal tonnage will increase by 73 percent; and
- Air cargo tonnage will quadruple.[3]

Although many freight facilities have experienced a decline in cargo volume due to the current recession, freight traffic is anticipated to continue to increase over the long-term as the U.S. population grows and consumes more goods. Increased demand for domestic and foreign goods is expected to result in the expansion of existing infrastructure or the development of new infrastructure to move freight faster and more reliably.

2.2 Air Pollution from Goods Movement

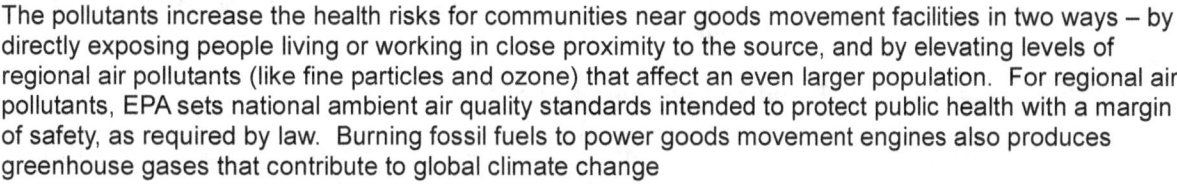

The ships, harbor craft, trucks, locomotives, aircraft and cargo handling equipment used to move goods in the U.S. typically rely on large, long-lived engines that burn diesel fuel (or similar fuels). These diesel engines emit soot particles and gases. Some of these gases are precursor compounds that can then react in the atmosphere with chemicals from other types of sources to form secondary air pollutants, like ozone and gaseous fine particles.

The pollutants increase the health risks for communities near goods movement facilities in two ways – by directly exposing people living or working in close proximity to the source, and by elevating levels of regional air pollutants (like fine particles and ozone) that affect an even larger population. For regional air pollutants, EPA sets national ambient air quality standards intended to protect public health with a margin of safety, as required by law. Burning fossil fuels to power goods movement engines also produces greenhouse gases that contribute to global climate change

---

[2] Cannon, James. U.S. Container Ports and Air Pollution: A Perfect Storm. 2008
[3] U.S. Federal Highway Administration. 2007 Freight Facts and Figures.
http://www.ops.fhwa.dot.gov/freight/freight_analysis/nat_freight_stats/docs/07factsfigures/pdf/fff2007.pdf

Emissions from diesel engines are complex mixtures consisting of a wide range of compounds including: directly emitted organic and black carbon, toxic metals, and other particulate matter (PM), plus gases like nitrogen oxides (NOx), sulfur oxides (SOx), volatile organic compounds (VOC), carbon monoxide (CO), formaldehyde, acrolein, and polycyclic aromatic hydrocarbons (PAH). While there are numerous hazardous chemicals in diesel exhaust, this report will focus on the impacts from emissions of direct PM, NOx, and SOx, as well as the resulting ozone and fine particle pollution formed in the atmosphere.

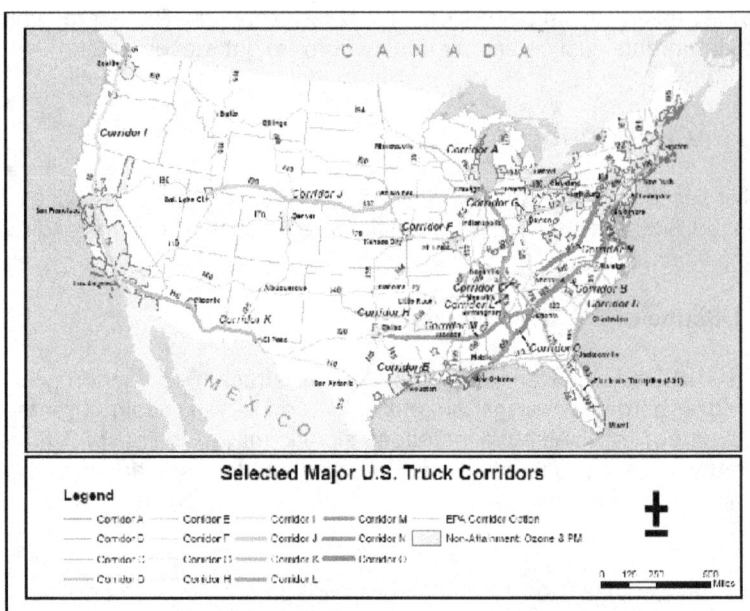

Non-Attainment Areas and Truck Routes

Particulate matter. Particulate matter is made up of tiny particles of solid or liquid suspended in a gas. Very small particles are directly emitted as a by-product of incomplete fuel combustion in an engine, and larger particles result from brake and tire wear. Diesel PM consists of a "core" of black elemental carbon with a coating of organic material and sulfates. Fine particulate matter (2.5 microns or less in diameter), known as PM2.5, includes directly emitted PM plus gaseous particles formed in the atmosphere from emissions of NOx or SOx and ammonia. EPA sets ambient air quality standards for PM2.5, as well as the coarser PM10 that is dominated by dust. California also regulates the subset of diesel PM as an air toxic.

Nitrogen Oxides and Sulfur Oxides. Emissions of both NOx and SOx can be directly associated with health effects or act as precursors for other secondary pollutants formed in the atmosphere. NOx compounds contribute to formation of both ozone and PM2.5. NOx reacts with ammonia, moisture, and other compounds to form nitric acid and related particles. In California, ammonium nitrate from NOx is the primary constituent of fine particles in the South Coast and San Joaquin Valley, which experience severe PM2.5 pollution levels. In the Eastern U.S., SOx is the more significant contributor to secondary PM2.5 levels. NOx can also react with VOCs to create ground-level ozone in the presence of sunlight. EPA establishes ambient air quality standards for ozone, nitrogen dioxide (NO2) and sulfur dioxide (SO2).

Air Toxics. Diesel exhaust includes more than 40 substances that are listed as hazardous air pollutants by EPA and are considered "cancer causing" by the California Environmental Protection Agency (CalEPA). Air toxics are chemicals known or suspected to cause cancer or other serious health effects, such as reproductive mutations or birth defects.

2.3 Health Impacts Due to Air Pollution from Goods Movement [4]

Diesel and other emissions from port and goods movement activities have significant human health and environmental impacts in onshore communities. These impacts include increased cancer rates, asthma, other respiratory and cardiovascular diseases, and premature death. Port and goods movement emissions also contribute to the formation of ground level ozone. Diesel engines at ports, rail yards and along truck routes create emissions that affect the health of workers and people living in nearby

---

[4] The first two paragraphs below are taken verbatim from: U.S. EPA Inspector General, *EPA Needs to Improve Its Efforts to Reduce Air Emissions at U.S. Ports, 09-P-0125,* March 23, 2009
http://www.epa.gov/oig/reports/2009/20090323-09-P-0125.pdf

communities, and contribute significantly to regional air pollution. EPA has determined that diesel exhaust is "likely to be carcinogenic to humans by inhalation" and that this hazard applies to environmental exposures.[5]

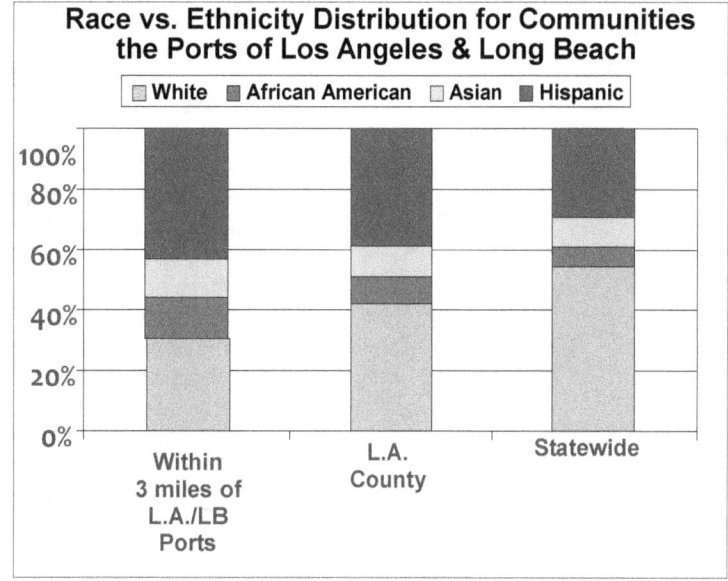

Recent studies show that populations living near large diesel emission sources such as major roadways,[6] rail yards, and ports[7] are likely to experience greater diesel exhaust exposure levels than the overall U.S. population, exposing them to greater health risk. For example, according to the California Air Resources Board, nearly 60 percent of the 2 million people living in the area around the Ports of Los Angeles and Long Beach have a potential cancer risk of greater than 100 in 1 million (due in part to port emissions), while over 410,000 people living closest to the same ports have a cancer risk greater than 200 in 1 million.[8] These cancer risk calculations are based on a unit risk value for diesel particulate adopted by the California Air Resources Board (CARB).

A significant body of peer-reviewed research studies now shows that air pollutants are higher in close proximity to mobile sources, such as highways.[9,10] A report issued by the Health Effects Institute (HEI) in May 2009 concluded that: *"Traffic-related pollutants impact ambient air quality on a broad spatial scale, ranging from roadside to urban to regional background. Based on a synthesis of the best available evidence, we identified an exposure zone within a range of up to 300 to 500 meters from a major road as the area most highly affected by traffic emissions."*[11] Several peer-reviewed articles have summarized the evidence about health effects in proximity to traffic-related air pollution[12,13,14] including studies showing an

---

[5] U.S. EPA (2002). *Health Assessment Document for Diesel Engine Exhaust*, prepared by the National Center for Environmental Assessment, Washington, DC, for OTAQ; EPA/600/8-90/057F.
[6] Kinnee, E. J., J.S. Touman, R. Mason, J. Thurman, A. Beidler, C. Bailey, R. Cook. *Allocation of on-road mobile emissions to road segments for air toxics modeling in an urban area.* Transport. Res. Part D 9: 139150, 2004.
[7] California Air Resources Board (CARB), *Roseville Rail Yard Study*, October 14, 2004; and CARB, *Diesel Particulate Matter Exposure Assessment Study for the Ports of Los Angeles and Long Beach*, April 2006. See: http://www.arb.ca.gov/railyard/hra/hra.htm
[8] California Air Resources Board (CARB), *Roseville Rail Yard Study*, October 14, 2004; and CARB, *Diesel Particulate Matter Exposure Assessment Study for the Ports of Los Angeles and Long Beach*, April 2006. See: http://www.arb.ca.gov/railyard/hra/hra.htm

[9] Zhu Y, Hinds WC, Seongheon K et al. Study of ultrafine particles near a major highway with heavy-duty diesel traffic. Atmos Envir 36 (2002) 4323–4335
[10] Greco SL, Wilson AM, Hanna SR et al. Factors influencing mobile source particulate matter emissions-to-exposure relationships in the Boston urban area. Environ. Sci. Technol. 2007, 41, 7675-7682.
[11] Health Effects Institute. Traffic-Related Air Pollution: A Critical Review of the Literature on Emissions, Exposure, and Health Effects. Special Report #17, 2009. Available at: http://pubs.healtheffects.org/view.php?id=306.
[12] Boothe ,V.L.; Shendell, D.G. "Potential health effects associated with residential proximity to freeways and primary roads: review of scientific literature, 1999 – 2006, Journal of Environmental Health. 2008, 70(8): 33-41

Reducing Air Emissions Associated With Goods Movement
A NEJAC Report of Advice and Recommendations

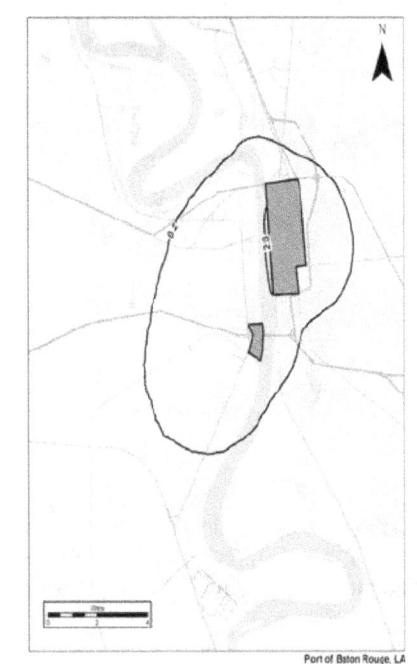

Concentration Isopleths for Baton Rouge, LA

Port of Baton Rouge, LA

increase in asthma and reduced lung function among children living in close proximity to traffic-related pollution.[15,16] With regard to health effects, the HEI Report concluded that *"Evidence was "sufficient" to infer a causal relationship between exposure to traffic-related air pollution and exacerbation of asthma and "suggestive" to infer a causal relationship with onset of childhood asthma, non-asthma respiratory symptoms, impaired lung function, and total and cardiovascular mortality."* In the same HEI Report, the HEI writers pointed out that, "Our conclusions have to be considered in the context of the progress made to reduce emissions from motor vehicles. Since the epidemiologic studies are based on past estimates of exposure, they may not provide an accurate guide to estimating health associations in the future."[17]

2.4 Community Impacts and Environmental Justice

As described above, good movement-related activities can have negative impacts on air quality and public health. Adjacent communities bear the burden of such activities resulting from the growth and demand for goods. Across the country, there are many communities near goods movement infrastructure that consist of large populations of low-income and minority residents. These environmental justice communities tend to have greater exposure to poor air quality as a result of diesel emissions from transportation facilities with high traffic density[18,19] This increased exposure may result in higher incidences of the health impacts described above in Section 2.3 among low-income and minority residents. As shown in Figure 1 to the right, the communities closest to the ports in southern California have a higher percentage of minority residents.

More recently, CARB released several additional rail yard health risk assessments, which all show that diesel PM emissions (from trucks, locomotives and yard equipment) result in higher risks of lung cancer in

---

[13] Salam, M.T.; Islam, T.; Gilliland, F.D. (2008). Recent evidence for adverse effects of residential proximity to traffic sources on asthma. Current Opinions in Pulmonary Medicine 14: 3-8.
[14] HEI. Traffic-Related Air Pollution: A Critical Review of the Literature on Emissions, Exposure, and Health Effects. Special Report #17, 2009.
[15] Gauderman W, Vora H, McConnell R, et al. The Effect of Exposure to Traffic on Lung Development from 10 to 18 Years of Age. Lancet 2007; 367:571-77
[16] McConnell R, Berhane K, Yao L et al. Traffic, susceptibility, and childhood asthma. Environ Health Perspectives. 2006 May; 114(5): 766-72.
[17] HEI. Traffic-Related Air Pollution: A Critical Review of the Literature on Emissions, Exposure, and Health Effects. Special Report #17, 2009, page 7-25.
[18] U.S. EPA. Control of Hazardous Air Pollutants from Mobile Sources: Regulatory Impact Analysis. February 2007
[19] Environmental Protection Agency. Control of Emissions of Air Pollution From Locomotive Engines and Marine Compression-Ignition Engines Less Than 30 Liters per Cylinder. Federal Register: June 30, 2008 (Volume 73, Number 126). Page 37100.

Reducing Air Emissions Associated With Goods Movement
A NEJAC Report of Advice and Recommendations

nearby communities.[20] The highest cancer risk was found among residents living across the street from a rail yard in San Bernardino, estimated at a risk of 3,000 out of one million (based on a 70-year exposure). The City of San Bernardino has a population of over 185,000 residents of which approximately 28 percent live below the Federal poverty level compared to the national individual poverty rate of 12 percent. In addition, the median family income in the city is $16,689 less than the national average of $50,046 according to the 2000 Census.

EPA recently analyzed a representative selection of national marine port areas and rail yards in order to understand the populations that are exposed to diesel emissions from these facilities,[21] using geographic information system (GIS) tools and census information.[22] The analysis showed, for example, that in Chicago the population living adjacent to the Barr Rail Yard, which has the greatest exposure to diesel emissions from that yard, is 97 percent African-American, while the general metropolitan area of Chicago is only 18 percent African-American.[23]

The EPA analysis shows that – across the country – the populations near major goods movement facilities are often minority and low-income communities.[24] Goods movement facilities may also be located near other industrial facilities and thus may contribute to existing local air quality problems. For example, in Houston, Texas, more than 20 percent of the area's largest industrial emission sources are located in East Houston, where the Port of Houston and the shipping channel that feeds it are located. Additionally, four major highways intersect this area resulting in high traffic density and additional air pollutant emissions. East Houston neighborhoods, which are predominantly minority and low-income communities, have the highest concentrations of air pollutants in Houston.[25] In California, Mira Loma, San Bernardino, Wilmington, Long Beach, Commerce, and Oakland are examples of environmental justice communities that are affected by emissions generated from marine port and locomotive related activities, distribution centers, and other transportation facilities associated with freight hubs.

The environmental, public health, and quality-of-life impacts of goods movement activities on communities are more pronounced in areas with major transportation hubs and heavily trafficked roads. Local areas with elevated levels of air pollution are of great concern to EPA and other environmental health agencies. The research described above shows that minority and low-income communities living near transportation hubs bear a disproportionate share of the environmental impacts because of their close proximity to multiple pollution sources.

2.5 Legal and Regulatory Environment

Air Quality. Under the federal Clean Air Act, EPA regulates air quality in the U.S. through the establishment of national ambient air quality standards for certain pollutants, including ozone, particulate matter, nitrogen dioxide, sulfur dioxide, and carbon monoxide. Regions that record air pollution levels above these standards are called "nonattainment" areas. States with designated nonattainment areas must prepare air quality plans, or State Implementation Plans (SIP), that identify the emission reductions needed to attain the standards and the control measures that will achieve those reductions.

---

[20] These studies are available at http://www.arb.ca.gov/railyard/hra/hra.htm
[21] ICF International. September 28, 2007. Estimation of diesel particulate matter concentration isopleths for marine harbor areas and rail yards. Memorandum to EPA under Work Assignment Number 0-3, Contract Number EP-C-06-094. This memo is available in Docket EPA-HQ-OAR-2003-0190.
[22] The Agency selected a representative sample of the top 150 U.S. ports including coastal, inland, and Great Lake ports. In selecting a sample of rail yards the Agency identified a subset from the hundreds of rail yards operated by Class I Railroads.
[23] ICF International. September 28, 2007. Estimation of diesel particulate matter concentration isopleths for marine harbor areas and rail yards. Appendix H. Memorandum to EPA under Work Assignment Number 0-3, Contract Number EP-C-06-094. This memo is available in Docket EPA-HQ-OAR-2003-0190.
[24] Ibid.
[25] See http://www.epa.gov/ttn/chief/conference/ei16/session6/bethel.pdf

A Federal Advisory Committee to the U.S. Environmental Protection Agency

A 2009 report by EPA's Office of the Inspector General noted that 31 U.S. seaports were located in nonattainment areas for ozone, PM2.5 or both, and projected that number would rise once EPA designated additional counties as nonattainment for the more health-protective 8-hour ozone standard established in March 2008.[26]

EPA and other air agencies have legal obligations to show continued progress in reducing air pollution emissions to meet ambient air quality standards, reduce exposure to air toxics, and achieve other health goals. Regulatory actions to control emissions from pollution sources provide the foundation for this progress, supplemented with voluntary initiatives.

Legal Authority and Recent Progress. Regulatory oversight of air emissions from goods movement sources is divided among international, national, tribal, state, regional, and local agencies. Typically, each agency enforces its own regulations and standards. Agencies at multiple levels may choose to share enforcement responsibility to increase monitoring or field inspections with the goal of improving compliance.

<u>International.</u> At the international level, ocean going ships (including foreign flagged) are subject to the rules of the International Maritime Organization (IMO) and its International Convention on the Prevention of Pollution from Ships. The U.S. Coast Guard serves as the lead agency for the U.S. delegation to the IMO. Representatives from EPA are invited to attend as part of the delegation. In October 2008, the IMO adopted tighter standards for ship engines and their fuels, set to phase in over the next decade. In regions with severe pollution problems, these requirements can be accelerated through establishment of Emission Control Areas (ECA) by the IMO. EPA has applied to the IMO for an ECA designation for the U.S. and Canadian coasts in cooperation with Environment Canada[27].

International aircraft are regulated by the International Civil Aviation Organization, with the U.S. Federal Aviation Administration leading the U.S. effort. International truck movements are subject to oversight under specific cross-national border initiatives or state actions to ensure international trucks meet U.S. emission standards.

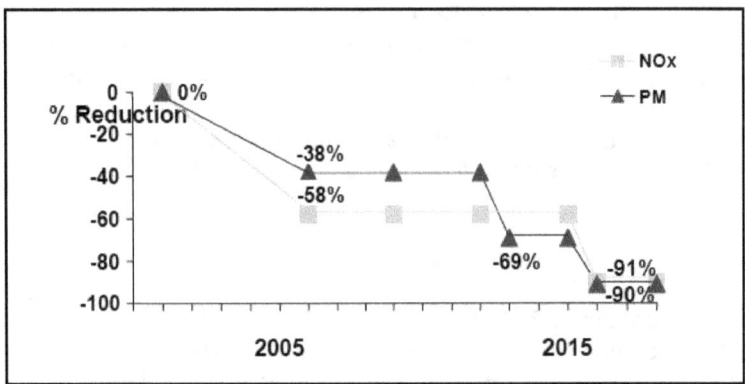

EPA Line-Haul Locomotive Standards: Reductions from Uncontrolled Levels

---

[26] U.S. EPA Inspector General, *EPA Needs to Improve Its Efforts to Reduce Air Emissions at U.S. Ports*, 09-P-0125, March 23, 2009 http://www.epa.gov/oig/reports/2009/20090323-09-P-0125.pdf
[27] For additional information about the ECA application to the IMO, see http://www.epa.gov/otaq/regs/nonroad/marine/ci/420f09015.htm

Photo by Andrea Hricko, USC

National. The federal Clean Air Act (CAA) and the National Environmental Policy Act (NEPA) provide legal authority to regulate and mitigate the impacts of emissions from goods movement in the U.S.

At the national level, EPA (in consultation with other federal agencies) is responsible for regulating emissions from trucks, locomotives, harbor craft, yard equipment, marine vessels and harbor craft, aircraft, and fuels under the CAA. EPA has promulgated mobile source regulations, including more stringent tailpipe emissions standards for new equipment (like trucks, locomotives, harbor craft, and cargo equipment) and requirements for the use of cleaner fuels, among other actions. Progress in reducing diesel emissions will be substantial over the next decade as new and rebuilt engines are introduced. However, the long life of these engines means that old, high-emitting, less-efficient technology will continue to operate for years to come.

By providing information, incentives, and financial assistance, EPA is working to encourage firms to adopt clean technologies that meet or surpass regulatory standards. The National Clean Diesel Campaign is an umbrella initiative that aims to reduce diesel emissions from various sectors, including trucks, locomotives, ships, and cargo handling equipment. EPA's Sector Strategy Program also works with industry to achieve sector-wide environmental goals. For example, EPA has encouraged ports to measure their environmental impact with emissions inventories and to deploy environmental management systems (EMS).

Freight transportation planning and infrastructure development are regulated by various departments of the U.S. Department of Transportation (DOT) including the Federal Highway Administration (FHWA), the Federal Railroad Administration (FRA), and the Federal Aviation Administration (FAA); as well as the Army Corps of Engineers and the U.S. Coast Guard.

The CAA imposes requirements on transportation planners. It requires that federally-funded or approved highway, seaport, airport, and rail projects conform to SIP emission projections to avoid creating new air quality violations, worsening existing violations, or delaying timely attainment of air quality standards. Federally-funded transportation projects that may generate significant traffic volumes are also required to perform a "hot-spot analyses" of PM10 and PM2.5 emissions. EPA partners with other federal agencies to set conformity policy via regulations and to enforce that policy as infrastructure proposals are approved.

Under the NEPA statute, Federal agencies (as well as those that receive Federal funding), must conduct a review of all potential impacts to human health and the environment resulting from a major Federal action. The environmental review must evaluate the action's direct and cumulative environmental impacts. NEPA also outlines a public involvement process for local communities to ensure that the health impacts of goods movement projects are properly considered and mitigation efforts are implemented. To further improve community involvement, EPA developed the *Final Guidance for Incorporating Environmental Justice Concerns in EPA's NEPA Compliance Analyses* to educate Federal agencies on ways to address environmental justice concerns and involve local communities.[28]

State/Tribal. States also play an important role in regulating goods movement through numerous other state environmental and transportation planning mechanisms. These mechanisms include: emission standards for fuels, restrictions on truck idling, and limits on visible smoke from diesel equipment. For example, Massachusetts, like 14 other states, has a policy to limit truck idling. Multiple states have also established smoke limits for big diesel engines.

The State of California can adopt more stringent emission standards for new engines or vehicles (subject to a waiver from EPA) and set fuel specifications. Other states can choose to opt into California rules, impose operational restrictions, and establish their own requirements to accelerate the turnover of existing equipment to cleaner models. In some states, legislation may be required to enable these actions.

California has an extensive program to assess and cut the health risk from goods movement sources, as well as to reduce the emissions that contribute to high regional ozone and PM2.5 levels.[29] CARB adopted rules requiring that existing diesel trucks, harbor craft, and cargo equipment be upgraded or replaced on an accelerated schedule. CARB rules also require the use of low sulfur fuel for ships ahead of the IMO requirements and use of shore-based electrical power (or equivalent alternatives) to cut ship emissions at dock.

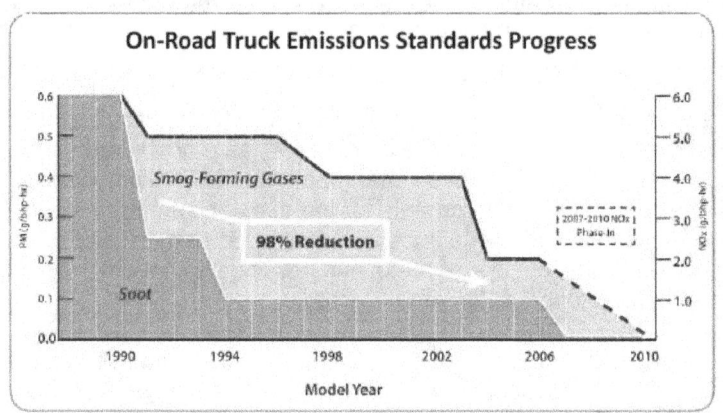

Source: U.S. Environmental Protection Agency, On-Highway Heavy Duty Diesel Emissions Reductions

Local. Local agencies, including ports and quasi-governmental organizations, can play a role in reducing emissions and health risk from freight facilities through their management of transportation corridors, and their zoning authorities affecting the location of freight infrastructure, and their use of landlord authorities to encourage or compel their tenants to transition to cleaner equipment and practices.

A number of ports have voluntarily implemented plans to manage air quality and reduce their environmental footprint. Ports have implemented a range of strategies, such as requiring shore power, increasing access to rail, and using low sulfur fuels among other strategies.
For example, the Ports of Los Angeles and Long Beach adopted a San Pedro Bay Ports Clean Air Action Plan[30] that calls for aggressive port action through leases, tariffs, and incentives to clean up diesel sources and limit the impacts of port expansion projects. The Port of New York and New Jersey has implemented a variety of clean air initiatives and has prepared a Harbor Air Management Plan.[31] Additionally, the Ports of Tacoma, Seattle, and Vancouver have developed the Northwest Ports Clean Air

---

[28] U.S. EPA http://www.epa.gov/compliance/resources/policies/ej/ej_guidance_nepa_epa0498.pdf
[29] CARB. http://www.arb.ca.gov/html/gmpr.htm
[30] San Pedro Bay Ports Clean Air Action Plan. www.cleanairactionplan.org
[31] IAPH Tool Box for Port Clean Air Programs. www.iaphworldports.org/toolbox%201/casestudies.htm

Strategy, which outlines a series of short- and long-term commitments for all facets of port-related emissions (i.e., ocean-going vessels, trucks, cargo handling equipment, rail, and harbor vessels).[32]

2.6 Land Use Planning and Zoning

Land use planning involves decisions about how land is used, including whether something gets built or expanded in a community, where it is built, and what concerns are addressed in the process. Land use planning decisions are often made at the local, regional, or state level. In most cases, the Federal government has limited authority to set policy in this arena. However, federally-funded projects, such as interstate highways and railroads, must comply with all Federal regulations including the General Conformity Rule of the Clean Air Act and Executive Order 12898, *Federal Action to Address Environmental Justice in Minority and Low-Income Populations*. As the lead agency in monitoring compliance of these policies, EPA can work to integrate environmental justice considerations into the planning and decision-making processes in order to mitigate burdens on minority and low-income residents.

Zoning and land use decisions at the local level affect the location of freight terminals and port facilities. Local zoning, and, in a few cases, state laws, also determine how close residential neighborhoods can be built to these facilities or how close new facilities can be built to existing neighborhoods. The siting of homes near highways, rail yards near schools, and recreational facilities near seaports and airports, can result in increased localized exposure for residents and school children. Notification and the involvement of affected community residents in the land use planning process is critical to making decisions involving growth, development, transportation investments, and the future of a community.

Photo by Caltrans

Increasingly, both residents and government officials recognize that a land use decision made in one local area can have impacts (positive or negative) on a broader geographical area, and that regional or system-wide impacts need to be considered when making such decisions. One example is the expansion of a marine port terminal that enables additional volume of containers imported from Asia. With that expansion, comes a need for more trucks and trains to transport the additional containers, coupled with a possible need for highway and rail yard expansion, and for larger warehouses to handle the influx of goods. Thus, a local decision made by a port authority near a city's harbor can have a regional effect in communities along transportation corridors, near rail yards, and in close proximity to distribution centers – even though these facilities may be nowhere near the port itself and may even be located in rural communities some hundreds of miles away. Conversely, not expanding the terminal might cause a different effect both in this and another region(s) since goods will continue to move from their origin to their destination by some form of goods transport, regardless of the expansion of an individual marine port terminal. Hence, it is critical to evaluate both localized and regional impacts prior to the initiation or expansion of goods movement related activities, to avoid and minimize the related health impacts and environmental justice concerns.

---

[32] Northwest Ports Clean Air Strategy.
http://www.portseattle.org/downloads/community/environment/NWCleanAirStrat_200712.pdf

The issues discussed above provide some background and context for a consideration of goods movement, air quality and public health, and environmental justice. The following sections in this report describe the recommendations of the NEJAC (based on the input from its Goods Movement Work Group) for further EPA actions to reduce the impacts of goods movement-related air pollution on environmental justice communities.

3. FINDINGS AND RECOMMENDATIONS

Consistent with the charge to the NEJAC from EPA, this section recommends ways in which EPA can work with its partners and stakeholders at the national, state, tribal, and local levels to reduce the risks to vulnerable communities exposed to goods movement emissions. The

Children playing at elementary school with truck expressway in background *Photo by Andrea Hricko, USC*

emission standards established by government agencies ensure that the diesel engines used to transport freight will ultimately be replaced with substantially cleaner models over the coming decades. The contribution of existing freight operations to elevated health risks in nearby neighborhoods and to high levels of regional air pollution, together with projected cargo growth over the longer-term, creates a need for additional actions now.

Recommendations for action in seven different focus areas are described below. These recommendations include both those things that EPA can directly influence, such as regulatory and enforcement mechanisms, as well as those arenas where EPA may play an indirect role, such as advocating with other agencies or encouraging voluntary implementation of cleaner technology and buffer zones. Within each area, the report describes the general principles and framework for taking action, followed by specific recommendations.

3.1 EFFECTIVE COMMUNITY ENGAGEMENT

The engagement of impacted communities may be initiated by residents and their representative organizations or by local, state, or Federal agencies involved in goods movement activities. Impacted community and tribal representatives seeking access to new or existing goods movements' activities can include different forms of "community facilitated strategies" or CFS. "Collaborative governance (CG)" is a complementary process that can effectively engage the community in collaborative decision making involving multiple stakeholders. CG typically is initiated by the executive branch of the government. CFS support robust empowerment and capacity building of the "community voice" necessary to reach across cultural, language, socio-economic, and technological barriers. CFS and CG should follow well-established moral and ethical principles that ensure transparency and accountability in dealing with prioritized goods movement activities.

There are two areas of consideration when addressing community driven engagement:

1. Improving traditional public participation mechanisms and procedures prescribed by law; and
2. Going beyond legal requirements using community driven mechanisms (such as a community facilitated strategy or a collaborative governance process) that incorporate the principles of environmental justice.

In addressing these two areas, it is important to identify and disseminate best practices that build upon EPA's previous work in effective public participation practices. Decision-makers and impacted communities can utilize various approaches and tools to ensure meaningful public participation that goes above and beyond the minimum legal requirements. Both approaches described below can advance these best practices and approaches.

3.1a. Community Facilitated Strategies. Involving the public, especially disadvantaged communities, in the decision making process is critical to achieving environmental justice for the communities and tribal territories likely to be impacted by proposed and existing goods movement related activities. This decision-making includes, but is not limited to, agreements, development of incentive programs, and other interactions and policy activities. Unfortunately, there are numerous examples in the goods movement sector where community involvement and meaningful public participation are lacking in the decision-making process. For example, while some port authorities webcast their harbor commission meetings (where decisions about new terminals or expanded operations are made), others do not post the meeting agendas online or have methods for making minutes publicly available. Some government transportation agencies have failed to consider impacted residents as stakeholders in the planning of highway expansions to accommodate increased goods movement activities until long after initial decisions were made. Anger and frustration has grown among residents who live near highways and transportation corridors when excluded from effectively participating in the funding, planning, and decision-making process.

Recognizing that every community and situation is different, an effective community-facilitated strategy would include at a minimum these elements at the local impacted areas[33]:

a. A community would determine through its own process the structure and form of a process to engage stakeholders in deciding on the best set of strategies to address impacts from existing or proposed goods movement facilities, infrastructure or activities.
b. The process would be convened by community or tribal leaders and assisted by legal, research, technical, and other groups that represent community and tribal interests.
c. Participants would be selected by the conveners. Potential participants would include community members, local businesses, government agencies, etc. Agencies which provide housing and related services for the homeless are also recognized as representative segments of impacted communities and tribal territories.
d. Participants would have access to independent technical and scientific expertise in order to understand impacts from facilities and activities, including cumulative and life cycle impacts.
e. All participants would have equal access to information and an equal voice at the table. Participants could request conveners to add parties necessary to achieve agreements on ways to address the identified impacts.
f. Participants would attempt to achieve those CFS agreements and if unable to do so, the community would choose among different legal, political, or other collaborative tools to move forward. A collaborative governance approach might be another way to move necessary parties though agreement processes and achieve acceptable outcomes.

As an example, a community-facilitated strategy includes the "Community Peoples' Table" (see figure 1) where all parties are actively and equally engaged in the decision-making process -- each party has a representative voice and a seat at the proverbial table. In the context of goods movement issues, CFS would empower consensus building among all residents, including low-income and minority communities and Native Americans in indigenous territories. Impacted communities would continue to have access to remedies through the legal protection, privileges, and rights under Federal and related international/state/local policies, regulations, statutes, and treaties.

---

[33] Community Perspective: "The West End Revitalization Association (WERA)'s Right to Basic Amenities Movement: Voice and Language of Ownership and Management of Public Health Solutions in Mebane, North Carolina" ; by Omega R. Wilson, Natasha G. Bumpass, Omari M. Wilson, and Marilyn H. Snipes: Progress in Community Health Partnerships Journal Fall 2008 • Vol 2.3, Page 237-243 (The Johns Hopkins University Press (pchp.press.jhu.edu)

Community Facilitated Strategy Paradigm

Community Facilitated Strategy Paradigm designed by Omega Wilson and submitted to NEJAC's Goods Movement Workgroup May 18, 2009

Principles and Framework: Residents have a right to voice their opinions and exercise their rights when a decision is going to impact them or their community. The "Community Peoples' Table" represents steps for building credibility and trust, for impacted areas, in a goods movement process that should foster transparency and accountability in policy and decision-making. Robust collaborative partnerships with impacted residents can help enlighten the decision-making process with community-based knowledge. The goal is to capitalize on existing community and tribal resources by building positive and effective working relationships between decision-making agencies and those adversely impacted by goods movement activities.

Key principles guiding action in this area are:

- **Affected communities should be fully engaged at the local, regional, and national level, during the planning, development, and implementation stages of goods movement-related decisions.** Efforts to engage and inform the impacted areas should begin early and continue through completion of the project or initiative. EPA can play an active role in ensuring that impacted communities and tribal territories are involved throughout the process. It is critical that the communities determine for themselves the structure and form of its engagement with other stakeholders. One example of the use of a community facilitated strategy was developed by the West End Revitalization Association (WERA) and residents of Mebane, NC when they were excluded for 16 years from the planning process of 8-lane corridor for a 27-mile bypass/interstate through two historic African American communities. Local and state transportation agencies designed the project with Federal funding that now has input from impacted property owners.[34]
- **Funding must be provided to plan, strategize, and implement actions at the community and tribal levels to mitigate health and environmental impacts from goods movement.** Equity in funding and parity in the management of collaborative problem-solving initiatives at the community and tribal level will ensure short and long-term measurable outcomes and sustainability.

---

[34] Work-In-Progress & Lessons Learned: "Use of EPA Collaborative Problem-Solving Model to Obtain Environmental Justice in North Carolina"; by Sacoby Wilson, Omega Wilson, Christopher Heaney, John Cooper; Progress in Community Health Partnerships Journal Winter 2007 • Vol 1.4 Page 327-337 (The Johns Hopkins University Press (pchp.press.jhu.edu)

- **The "community voice" is recognized as valid and important to the resolution of goods movement issues that include air quality hazards as well as related water and soil risks.** Long-time community and tribal members can provide valuable information, such as the location of a tribal burial ground, past uses of sites, historic sites, and undocumented hazards that can positively influence goods movement planning and mitigation decisions.
- **Consideration of the cumulative and multiple impacts of all aspects of the goods movement supply chain (from mining of raw materials and manufacturing to landfill disposal and recycling) is an important part of the meaningful community involvement in resolving concerns about the impacts of goods movement activities.**

Recommendations: Based on the above mentioned principles, the following identifies specific recommendations for EPA action to directly effect, or influence, the needed changes.

1. EPA should promote decision-making processes that empower impacted community and tribal stakeholders through collaborative problem-solving approaches, that include:
    - Implementing a comprehensive outreach strategy by which to deploy the use of community facilitated strategies in communities where goods movement operations have been identified by EPA as high priority (*see complementary recommendation in section 3.3 – Regulatory and Enforcement Mechanisms*). Such a strategy must be transparent and accountable. It will also ensure that community stakeholders are included in advisory, planning, and decision-making,
    - Implementing new policies that support community-owned and -managed research data within impacted communities and tribal areas, and include social, economic, cultural, and community health factors.[35]
    - Evaluating and updating its EPA public participation approaches related to their effectiveness within communities affected by goods movement activities. A starting point would the updated recommendations put forth by the NEJAC in its Model Plan for Public Participation (1994).
    - EPA should encourage other federal agencies to adopt these recommendations.
    - Taking the lead in evaluating and validating the "community voice" and promoting a shift towards community-based approaches to capacity building, funding, and collaborative problem-solving.[36]
2. EPA should ensure that sustainable resources are available to increase the capacity of the community- and tribal-based organizations to participate in both traditional public participation processes and CFSs from within impacted communities and tribal territories. Community capacity includes the ability to document community-driven data collection, produce reports of results, and present evidence in informed manner, with the assistance of legal, research, and technical experts. Some examples include workshops and trainings for the CFS participants about related issues. These resources should be monitored to ensure the sustainability of funding equity and management parity for community and tribal based environmental justice organizations.
3. EPA should engage environmental justice areas and their locally based organizations to prioritize goods movement activities and related risks using the community facilitated strategy as a tool to address site-specific concerns. Human exposures, health effects and care as well as risks to impacted stakeholders' residential, business, and public properties should be among those priorities and concerns.

3.1b Collaborative Governance and Problem-Solving Strategies

Collaborative governance is a term that describes a shared decision making process involving representatives from the public, private and non-profit sectors, citizens, and others. These individuals

---

[35] Theory and Methods: "The West End Revitalization Association's Community-Owned and -Managed Research Model: Development, Implementation, and Action"; by Christopher D. Heaney, Sacoby Wilson, and Omega R. Wilson; Progress in Community Health Partnerships Journal Winter 2007 • Vol 1.4 Page 339-349 (The Johns Hopkins University Press (pchp.press.jhu.edu)

[36] "Built Environment Issues in Unserved and Underserved African-American Neighborhoods in North Carolina"; Sacoby M. Wilson, Christopher D. Heaney, John Cooper, and Omega Wilson; Environmental Justice Journal, Volume 1, Number 2, 2008, Page 63-72 (Mary Ann Liebert, Inc. Publisher)

may be able to contribute knowledge or resources, in developing effective, lasting solutions to public problems that go beyond what any sector could achieve on its own. It has been used to address many complex public issues and is well suited to address many environmental justice issues. Collaborative governance takes as its starting point the idea that working together creates more lasting, effective solutions. It is one of the tools that might be invoked by the participants in a CFS process to tap into additional means of financing desired investments or leveraging other resources.

In cases where a community facilitated strategy has been successfully implemented, all the needed participants will be at the table and will have agreed to take action, including official decisions, to support an agreed upon strategy. In other cases, for any number of reasons, not all the needed parties will have come to the table and further process will be needed to implement the solutions agreed to at the People's table. In these cases, and others where the impacted stakeholder community and tribal leaders believe they already possess sufficient capacity and resources to engage in collaborative decision-making, or where it is desired to bring additional resources to bear to implement one or more solutions, it may be suitable to invoke a collaborative governance process.

For those situations for which it is decided that a collaborative governance approach would be helpful, community leaders, together with other participants in their process, would first request an agency, foundation, civic organization, or public-private coalition, to act as a sponsor in providing funding and other support for a collaborative governance process. The community, with the sponsor, would engage an impartial organization to perform an assessment to determine the likelihood of success for the process by talking to all potential participants. If the assessment is favorable, the community and others who support the process would request a governor, legislator, local official, or respected civic leader to act as a convener -- with power to bring diverse people together in order to reach agreement on needed solutions and how they will be implemented. The convener could, if requested, appoint one of the leaders of the community facilitated strategy to become a co-convener of the process. The impartial organization would assist the conveners to identify all the appropriate participants and ensure skilled process management. At the end of the process, all the participants would enter into an agreement committing them to implement the agreed upon solutions.

An environmental justice use of a collaborative governance system was in the North Portland Diesel Emissions Reduction project, which was initiated at the request of a local environmental justice organization. The Governor of Oregon appointed a local convener who, with the assistance of the National Policy Consensus Center (a neutral organization), brought together government agencies and private and public trucking fleets, which all agreed to reduce diesel emissions through fuel and equipment upgrade projects. Financing for projects and actions agreed upon was shared by public and private entities to support the stakeholders' voluntary commitments. The West Oakland Environmental Indicators Project was co-convened by a community organization and EPA. They negotiated a joint partnering agreement, which was open to all stakeholders. Among the initial results were conversion of a number of heavy-duty trucks to compressed natural gas; and an emissions reduction program for the Port of Oakland for about 2,000 trucks.

A collaborative governance approach to goods movement-related issues would likely enhance both the chances of achieving agreed upon solutions and the outcomes sought by the community. Similarly, if a community develops a series of strategies requiring coordinated and complementary actions by a variety of public and private entities; a collaborative governance process convened by a governor, a mayor, an agency and community together, or another respected leader, has an excellent chance of achieving multiple enhanced outcomes. These might include infrastructure design that reduces air quality impacts, relocation of playgrounds or schools, supplemental pollution controls and health monitoring

While much of the experience with collaborative governance approaches has been at the community level, there is also a need for them at a more regional level. Many of the decisions that have the potential to create or ameliorate environmental justice issues have been and will be made at metropolitan, multi-county, tribal and state/county, multi-state levels. They include investments in transportation infrastructure, the siting of controversial developments like distribution centers, policies on alternative fuels availability and many others. While participants will include a different set of actors, the empowered representation from impacted communities is essential. There also is a need for robust and innovative stakeholder alliances, partnerships, and collaborative governance approaches to foster solutions to environmental concerns at a more regional level

Principles and Framework. Collaborative governance mechanisms may be appropriate in some circumstances to assist in implementing needed emissions reductions strategies for goods movement in communities with environmental justice issues. Key principles guiding action in this area are:

- During a collaborative governance process, all necessary groups, jurisdictions, and authorities, especially groups representing impacted communities, should have a meaningful part in making decisions at both the regional (multi-state, state, and multi-county/tribal) and at the community level (tribe, city, county, and neighborhood) on strategies and investments that will reduce air emissions associated with goods movement. To ensure the participation of groups from impacted communities, funding for their participation and for technical assistance needed to participate on an equitable basis should be assured early in the process, as would happen with a community facilitated strategy. To ensure the collaborative process is as objective and neutral as possible, community members must have a role in selection of the convener and the neutral process manager/facilitator.
- As part of collaborative governance processes, government participants should work together as an integrated group in order to promote an efficient process and resolve internal conflicts themselves.
- Collaborative governance processes should adhere to the principles of equity and inclusiveness; respect; transparency; effectiveness and efficiency; responsiveness; accountability; forum neutrality; and consensus-based decision-making. The goal is to reach agreements that might not be possible or as comprehensive using other means such as negotiation, mediation, settlements, or other forms of conflict resolution.
- Agreements reached through a collaborative governance process should aim to maximize beneficial outcomes and reduce costs across regulatory, technology, and other sectors. The goal will be to reach agreements on implementing integrated mitigation actions, including investments in infrastructure and decisions about land use, community benefits, local incentives, financing and funding mechanisms, job creation, and relocation.

Recommendations. Based on the principles described above, this section identifies specific recommendations for EPA action to directly effect, or influence, the needed changes:

4. EPA should support, encourage, and, where appropriate, co-fund collaborative governance processes relating to goods movement issues at both regional and community levels. Initially, EPA should co-fund several demonstration projects. EPA should seek commitments by federal and state agencies, regional organizations, municipalities, goods movement entities, foundations, and others to help fund these processes and the projects that are agreed upon. However funded or convened, these processes should assure that all appropriate participants should be included.
5. EPA should take the lead to get other Federal agencies to provide scientific and technical advice to these processes and to assist in implementing agreements. EPA should encourage all the participating Federal, state and local agencies to coordinate their authorities, technical assistance, and investments.
6. EPA should assist in identifying and supporting collaborative governance and consensus programs, private neutral facilitators, or equivalent experts to assist in process design, support to conveners, management, and facilitation. There is a network of mostly university-based centers that have experience both in traditional conflict resolution and in the emerging field of collaborative governance. These centers, as well as others that may be more conveniently located to the community, could serve as the neutral forum and provide process management and facilitation.

3.2 HEALTH RESEARCH DATA GAPS AND EDUCATION NEEDS

As noted in the Background section of this report, emissions from port, rail, trucking, and other goods movement activities have significant human health and environmental impacts in onshore communities. EPA has determined that diesel exhaust is "likely to be carcinogenic to humans by inhalation and that this hazard applies to environmental exposures." In addition, recent studies show that populations living near large diesel emission sources such as major roadways, rail yards, and ports are likely to experience greater diesel exhaust exposure levels than the overall U.S. population, exposing them to greater health risk[37].

Although there are research efforts to quantify goods movement-related emissions and resultant health effects, significant data gaps exist. Addressing these gaps would help to better elucidate community exposures and health effects. These could include air monitoring, development of emission inventories, exposure assessment studies, toxicologic and epidemiologic studies, health impact assessments (HIA), and health risk assessments (HRA).

SOME DIFFERENCES BETWEEN HEALTH RISK ASSESSMENTS AND HEALTH IMPACT ASSESSMENTS	
Health Risk Assessment (HRA) *	**Health Impact Assessment (HIA)**
Purpose: To quantify the health effects from a change in exposure to a particular hazard (*e.g. an air pollutant*).	**Purpose:** To make evidence based judgments on the health impacts of public and private decisions, and make recommendations to protect and promote health
Focus is primarily on one exposure-impact pathway (*e.g., an increase in diesel exposure leading to lung cancer*)	Takes a holistic approach to predict health outcomes of a variety of environmental and social impacts from a proposed project, program or policy; HIAs look at a range of exposures (including social exposures as well as environmental)
Predicts risk of health impact to a large population using estimations calculated from models	Can include qualitative (*e.g., surveys*) and quantitative (*e.g., modeling*) methods of analysis to evaluate potential health impacts
Does not directly measure pollutants (hazards) or exposures	Uses existing data and analysis when available, but primary data collection may be undertaken as needed
Does not include a plan for monitoring the impact of HRA or the proposed project on future health outcomes	HIA practice standards include monitoring as an important follow-up activity in the HIA process to track the outcomes of a decision and its implementation
May be included as part of an EIS, used as a tool for assessment in an HIA, or as a stand-alone assessment	May be included as part of an EIS, or as stand-alone assessment
	* as conducted in California for diesel cancer risks

HRAs have been conducted at ports and rail yards in California, looking at both cancer and non-cancer health outcomes, such as cardiovascular and respiratory illnesses and premature death.[38, 39] That state's Air Resources Board (CARB) has characterized the near-source impacts of diesel exhaust emissions

---

[37] U.S. EPA Inspector General, *EPA Needs to Improve Its Efforts to Reduce Air Emissions at U.S. Ports, 09-P-0125,* March 23, 2009 http://www.epa.gov/oig/reports/2009/20090323-09-P-0125.pdf
[38] California Air Resources Board (CARB), *Health Risk Assessments and Mitigation Measures for 18 Rail yards.* See: http://www.arb.ca.gov/railyard/hra/hra.htm
[39] CARB, *Diesel Particulate Matter Exposure Assessment Study for the Ports of Los Angeles and Long Beach,* April 2006. ftp://ftp.arb.ca.gov/carbis/msprog/offroad/marinevess/documents/portstudy0406.pdf See also HRA results for West Oakland at http://www.arb.ca.gov/ch/communities/ra/westoakland/documents/factsheet112508.pdf

because it has a unit risk value for diesel exhaust's cancer effects. EPA has considered but chose not adopt a unit risk value for diesel exhaust cancer effects. This gap is critical because the current characterization strategies used by EPA are based on fine particle measurements, and they do not necessarily reflect the full exposures or health risks in these communities, due to limitations in monitoring and modeling and to the fact that fine particle measurements do not adequately reflect near-traffic exposures.

In addition, building awareness of the potential impacts of goods movement activities is crucial to addressing environmental injustices. Numerous activities can be used to educate residents, community-based groups, elected officials and others about the potential impacts of emissions from goods movement facilities and transportation corridors. These include:

- Educational conferences and workshops to share research findings on health effects, community concerns, and workable solutions. An example would be "Moving Forward: a conference on healthy solutions to the impacts of goods movement[40]" in 2007, with 550 attendees from 16 states and four countries, which provided an opportunity to share research results, community concerns, and solutions with a network of scientists, regulators and community groups.
- Fact sheets, videos, or reports about the impacts of goods movement activities (and successful mitigation measures) at a state or local level, written by government agencies, nonprofit groups, or university researchers. Examples include "A View from Our Window[41]" and "Paying with Our Health,[42]" both produced by community-based coalitions, a "Goods Movement 101" curriculum,[43] and several articles in *Environmental Health Perspectives*[44].

Principles and Framework. In many cases, residents living near goods movement facilities and along transportation corridors are disproportionately impacted by ship, rail, and truck emissions. To better understand the magnitude of emissions from goods movement facilities and the potential health impacts from exposure while addressing data gaps, additional monitoring, research studies, analyses, and information campaigns are warranted. Key principles guiding action in this area are:

- There is a need for more near source/localized air pollution monitoring stations because central site monitors do not adequately reflect the higher levels of exposure to mobile source pollution that communities face in close proximity to goods movement facilities. Fine particle measurements (PM2.5) do not fully reflect the levels of diesel exhaust emissions to which residents are exposed.
- There is a need for EPA to review the current research findings on diesel exposure and cancer and the current status of its methods to characterize diesel risk.
- There is a need for emissions inventories and air pollution monitoring to better understand the magnitude of emissions at major goods movement facilities, hubs, and corridors. Emissions inventories have been completed for only a small number of highway expansion projects, major ports and rail yards and none have been developed for large distribution center complexes. Only a few major ports [45] and one major rail yard in California have air monitoring programs with results publicly available and to our knowledge no other goods movement facilities have air monitoring programs with results publicly available.

---

[40] See "Moving Forward Conference" at www.TheImpactProject.org
[41] See http://www.ccaej.org/docs/MAC/MAC_rev_12-8-05.pdf
[42] Pacific Institute. *Paying with Our Health: The Real Costs of Freight Transportation in California.* 2007. www.pacinst.org/reports/freight_transport/PayingWithOurHealth_Web.pdf
[43] See "Goods Movement 101" at http://www.TheImpactProject.org.
[44] Hricko, A. *Global Trade Comes Home: Community Impacts of Goods Movement.* Environmental Health Perspectives, February 2008. Available at http://www.ehponline.org/members/2008/116-2/spheres.html; Hricko, A. Guest editorial. *Ships, Trucks, and Trains: Effects of Goods Movement on Environmental Health.* April 2006. Available at http://www.ehponline.org/docs/2006/114-4/editorial.html
[45] See, for example, San Pedro Bay Ports (Port of Los Angeles and Port of Long Beach) Clean Air Action Plan, air monitoring program, http://caap.airsis.com/

- There is a need for additional scientific studies involving emissions from goods movement, including exposure assessment, toxicology, and cumulative impacts analysis, and health impact/risk studies. Although California has estimated diesel cancer risks and non-cancer health effects at 18 rail yards and a number of ports, no other state has conducted such studies. There is a need to build on existing occupational health studies with additional peer-reviewed studies of actual emission levels at, or health effects related to people living near ports, rail yards, or distribution centers.
- There is a lack of sufficient research funding across Federal agencies and research institutes to conduct studies of goods movement emissions and health impacts. Although EPA and the National Institute of Environmental Health Sciences (NIEHS) have partnered to fund Children's Environmental Health Centers, EPA and NIEHS do not have a similar partnership or special priority areas for funding research on the health impacts of goods movement. In addition, the U.S. Department of Transportation (DOT) does not have its own health research agenda in this area, nor a joint program with EPA and/or NIEHS.
- There is a lack of national attention and information/education about the issue of air pollution and health effects from goods movement facilities and freight/transportation corridors.

Recommendations. Based on the principles described above, this section identifies specific research and education recommendations for EPA to guide future action.

7. EPA should establish, for the port and rail sectors, a list of the largest ports and rail yards in the United States, and complete the analysis of demographics near port and rail facilities that was begun in conjunction with the 2007 Locomotive and Marine Engine Rule[46]. EPA should also undertake an assessment of the contribution from off-site transportation highways/corridors adjacent to those facilities (e.g., from trucks transporting goods from a port to a rail yard or distribution center). This will allow EPA to better understand the goods movement locations where significant environmental justice concerns may exist, even though community residents may not have raised concerns.
8. EPA should direct each Region to develop a plan to prioritize the most significant goods movement facilities of potential concern for emissions impacts within each region. The priority list should be based on emissions estimates from facilities and off-site transportation emissions, relative size of the facility, anticipated growth, proximity to disadvantaged communities, cumulative impacts, community concerns, and other relevant factors. Additionally, these priority lists should utilize information that already is available, such as emissions inventories, HRAs, action plans that have been developed to reduce emissions, air monitoring results, and scientific research results.
9. For those priority facilities, EPA should provide funding and technical guidance to state or local air agencies to conduct localized monitoring for toxic air pollutants in close proximity to the top priority goods movement hubs and corridors, with results available to the public.
10. EPA should conduct and/or fund additional research studies, including:

    - Studies of exposure assessment, emission characteristics of both on-site and off-site sources, and source apportionment studies of emissions from goods movement facilities, including research on coarse, fine and ultrafine particles
    - Toxicologic studies (e.g., animal and biomarker studies and assays);
    - Epidemiologic studies of health effects of residents or school children in communities impacted by goods movement.
    - Cumulative impacts studies

    As guidance in facilitating research and studies, EPA should review the list of research gaps in the HEI Report[47] on the health effects of traffic-related air pollution. EPA should consider developing a three-way funding partnership with NIH (NIEHS) and DOT (FHWA, FRA, and FAA) to fund research

---

[46] See Final Rule: Control of Emissions of Air Pollution from Locomotives and Marine Compression-Ignition Engines Less Than 30 Liters per Cylinder **(published May 6, 2008 and** republished June 30, 2008)

[47] See in particular, Section 5, part VIII.1 and Table 7.6 of the HEI Report, found at http://pubs.healtheffects.org/view.php?id=306

on exposure assessment, toxicologic, and epidemiologic studies related to exposure to emissions from the goods movement industry. The partnerships should include community-driven research and participation, including outreach and education.

11. EPA should revisit its health assessment of diesel exhaust emissions[48] as the Agency indicated it would do when it issued its assessment document in May 2002. Considering research that has occurred in the interim, and evaluating the need for further research, EPA should conduct a review of the current status of diesel risk characterization and the current scientific studies on diesel exhaust exposure and its links to cancer in order to determine if the Agency should reconsider adopting a unit risk value for diesel exhaust. In its scientific review, EPA should consider other health outcomes from exposure to diesel emissions, such as cardiovascular and respiratory illnesses.[49]

12. EPA should consider advocating that health impact assessments (HIA) or similar analytical assessments be conducted for major new or expanding goods movement facilities and transportation projects/corridors that are covered under NEPA. Some EPA Regional offices are already requesting that ports and freeway expansion projects conduct such HIAs, which are comprehensive health analyses of proposed infrastructure projects that evaluate air pollution, noise, impacts on access to parks, and other broad health-related issues.

13. EPA should develop a national communications plan to reach elected officials, urban planners, transportation officials and community members with information about the emissions from, and health impacts of, goods movement activities, using the same techniques the Agency has used in its SmartGrowth activities. Such a campaign should include fact sheets on each goods movement sector, in a number of languages, that summarizes concerns about emissions and health effects research findings. The information should be readily accessible on the EPA national and regional websites.

14. EPA should develop a special funding stream for environmental justice community grants focused on goods movement communities, to include community-based participatory research related to health impacts.

3.3 REGULATORY AND ENFORCEMENT MECHANISMS

There are numerous regulatory strategies to reduce freight emissions and exposure. These strategies include:

- Cleaner new engines and fuels for ships, harbor craft, locomotives, trucks, equipment, and aircraft;
- Fleet modernization to accelerate the replacement of existing diesel equipment with dramatically cleaner models, or to upgrade the existing equipment by: replacing the engine with a cleaner version ("repower") or installing additional verified pollution control devices ("retrofit");
- Shore-based electrical power for ships and harbor craft to eliminate engine operation while at dock.
- Operational limits on unnecessary idling for trucks, locomotives, and equipment;
- Low speed zones for ships to cut NO_x emissions that contribute to PM2.5 and ozone on-shore;
- Restrictions on visible smoke emissions from trucks, locomotives, or other sources; and
- Targeted enforcement actions for freight facilities in highly impacted communities.

In some cases, these strategies also have been successfully implemented or accelerated with a voluntary, collaborative basis through such means as enforceable agreements or incentives, etc. [*See section 3.7-Financing for additional discussion of these voluntary strategies*]

Principles and Framework. A regulatory approach historically has been the foundation to cut the impacts of freight movement on nearby communities, and to reduce regional pollution levels that can also affect environmental justice areas. Effective regulation of freight-related air pollution depends on action

---

[48] See http://cfpub.epa.gov/ncea/cfm/recordisplay.cfm?deid=29060
[49] See for example, a letter from Region 9 EPA to the Army Corps of Engineers concerning a Port of Long Beach marine terminal expansion project at http://www.epa.gov/region09/nepa/letters/Port-Long-Beach-Middle-Harbor-Redev-Proj.pdf

by each entity with relevant authority, including international bodies, federal agencies, tribes, states, and local agencies (including seaports and airports).

EPA must play multiple roles -- as a direct regulator and enforcer of federal requirements; as an advocate with other federal agencies about transportation policy, including mitigation and funding; as a strong supporter of aggressive international treaties; and as a facilitator of state and local initiatives that go beyond federal requirements. EPA and other government entities are engaging in each of these areas, but there are still impediments to the fast-paced progress needed to address the environmental justice concerns. Key principles guiding action are:

- There is a need for more urgency in national actions to cut freight pollution to speed attainment of air quality standards by the applicable deadlines and to reduce the exposure and health risk from freight emissions in communities that already attain those standards.
- The pace of fleet modernization must be accelerated through a combination of regulatory and incentive mechanisms.
- More vigorous, focused enforcement should be used to improve air quality in communities affected by goods movement facilities.
- Quantitative goals and policies are needed to cut criteria and toxic pollutants from existing, expanding, and new goods movement facilities.

Recommendations. EPA leadership should elevate the issue of environmental justice related to goods movement activities to initiate additional regulatory and enforcement strategies within the Agency, as well as to facilitate action by other Federal, state, tribal, and local agencies.

15. EPA should ensure effective, early control requirements on international ships and aircraft. On the marine side, EPA should work with neighboring countries to achieve IMO approval of a North American Emission Control Area (ECA) to accelerate deployment of new IMO standards for cleaner ships and fuels. EPA should work with FAA to introduce stringent proposals to the International Civil Aviation Organization for aircraft engines with lower NO_x and PM emissions, as well as cleaner jet fuels. Concurrently, EPA should publicly evaluate the potential benefits, costs, and impacts of pursuing new national regulations requiring advanced control technology and cleaner fuels for both U.S. and foreign-flagged ships operating in U.S. waters, and aircraft serving U.S. airports.
16. Significantly accelerate modernization of the existing diesel fleet used to transport freight. EPA should fully use its programmatic authorities to achieve additional, earlier reductions from existing goods movement sources. EPA should also encourage its federal partners to support these efforts through incentives and other mechanisms. EPA's actions should include, but not be limited to: Requiring or updating engine rebuild standards for all existing engines under its authority; Using all available means to encourage engine and equipment manufacturers to accelerate the development and production of the cleanest engines in advance of regulatory deadlines sources; and Evaluating and assessing operational opportunities to reduce in-use emissions, such as adopting a national, time-limited idling standard for all engines under its jurisdiction. (*See Financing section for complementary incentives element*)
17. EPA should facilitate state and local initiatives that go beyond Federal requirements to cut community and regional pollution. EPA's role should include:
    - Providing technical assistance to states that want to adopt and enforce in-use emission standards to accelerate fleet modernization, as allowed by federal and state law.
    - Issuing timely waivers for stricter California vehicle and fuel emission standards to benefit all states wishing to "opt-in" to those standards.
    - Supporting expansion of state/local operational restrictions, including but not limited to idling limits and designated truck routes, with information about successful programs that could serve as models.
    - Using Federal leverage (via project approval authority and funding capability) to aid state/local efforts on legal agreements with industry to accelerate progress (early availability of cleaner engines depends on recommendation 16 above).

18. Establish quantitative goals to reduce emissions and exposure from existing, major freight facilities and plans to achieve those goals. EPA, in consultation with states and communities, should identify sites of concern and establish priorities among them. EPA should employ available planning mechanisms to set such goals, either by identifying national targets or assisting local or state efforts. EPA and other Federal agencies should encourage ports, marine terminal operators, railroads, airports, and transportation agencies, etc., to develop freight facility air quality plans in a public process with: quantitative reduction goals; commitments for action to achieve those goals based on voluntary initiatives with public agency involvement, enforceable agreements, the facility's legal authorities, and/or incentives; and periodic public reporting on progress. With this mechanism, EPA and partner agencies can offer assurances to environmental justice communities regarding the magnitude and pace of emission reductions from high priority freight facilities. In nonattainment areas, EPA should back these goals with enforceable SIP commitments for future federal actions to reduce emissions from goods movement sources for timely attainment.

19. Mitigate localized air impacts from expanding existing freight facilities or siting new ones. If full mitigation is not feasible, EPA should establish policies and guidance to assure that new and expanded infrastructure and/or facility projects will achieve the highest technically feasible air levels and be mitigated to the extent acceptable to impacted neighborhoods. As part of the guidance, EPA should outline a process based on the principles and recommendations in Section 3.1 (Effective Community Engagement) of this report. To accomplish this, EPA should work with DOT agencies to require more effective general and transportation conformity programs to ensure that affected projects cannot simply use the expected reductions from other sectors to subsidize growth in operations. In communities already impacted by high pollution levels from freight facilities, expansion and new facilities should not be considered unless the project and its mitigation measures can be designed to at least "do no harm" to the localized area, as well as the region.

20. Expand enforcement. EPA should increase its enforcement efforts, in coordination with state/local authorities, by deploying more field inspection teams to focus on sources operating at goods movement facilities and within nearby communities. EPA should also target violation penalties to help fund fleet modernization by directing enforcement fines toward diesel clean up projects in environmental justice areas.

21. EPA should vigorously implement and enforce on-time implementation of all current mobile fleet clean fuel and emission reduction regulations.

3.4 LAND USE PLANNING AND ENVIRONMENTAL REVIEW

Improvements in land use planning are an important component of any overall strategy to prevent or mitigate the air emission and community impacts related to goods movement facilities. Consideration of existing sensitive receptor locations (such as schools, homes, hospitals, nursing homes, daycare centers) should be considered whenever siting new (or expanding existing) goods movement facilities. Similarly, the presence of existing goods movement facilities (ports, rail yards, truck traffic corridors, distribution centers) should be considered when siting facilities for sensitive receptors.

Through the transportation conformity process (in which a state SIP conforms to a State Transportation Plan), EPA has influence over air quality when new transportation corridors and/or federally-funded freight facilities are constructed or expanded. Also, states with non-attainment areas have to submit to US EPA a wide array of control measures considered in the development of State Implementation Plans (SIP). In a SIP, a land use measure might be a policy or program that changes the urban form in a way that leads to fewer vehicle emissions.

Principles and Framework. The construction or expansion of goods movement facilities, infrastructure, and transportation corridors has potential impacts on air quality, land use, and environmental justice. A key aspect of land use planning in goods movement communities involves providing zones of separation or buffers between new residential or school developments and port/freight facilities or between new or expanding freight hubs and existing communities and schools. After considering evidence regarding the health risks of air pollution (including diesel exhaust and other emissions), establishing zoning designations and other land use planning actions can help to ensure that new facilities and transportation corridors are constructed in a manner that minimizes future risks to surrounding communities and prevents siting of conflicting uses. Such actions will help ensure that sensitive receptors are located at a safe distance from goods movement transportation and infrastructure.

Because of the array of different actors involved in land use planning, there are numerous policy levers that EPA can use to achieve environmental justice goals. There are a number of key principles guiding action in this area:

- **Scientific studies show that adverse health impacts can be minimized with increased distance between sources of air pollution and sensitive receptors, with most studies showing that respiratory health effects (e.g., exacerbation of asthma) are more likely to occur within 300-500 meters of traffic-related air pollution.** These research findings should be used in developing land use guidelines. The Health Effects Institute (HEI) Report says: "In light of the large number of people residing within 300 to 500 meters from major roads *[that is, near traffic-related pollution]*, we conclude that the evidence for these *[adverse]* health outcomes indicates that the exposures are likely to be of public health concern and deserve public attention. *Italics added.* "Traffic-related pollution" results from trucks traveling on highways going to rail yards, ports and distribution centers and operating at those facilities, as well as diesel and other equipment whtissg operating at those goods movement facilities.
- **Land use decisions are among the most controversial urban planning issues for community residents, and early community involvement is critical to high quality environmental health decisions.** Improved public involvement can ensure that community concerns are addressed and that grass roots solutions to environmental problems are considered. Improved public involvement and collaborative approaches can ensure that community needs, economic development, and other concerns are addressed; grassroots solutions to environmental problems are considered; and multi-jurisdictional planning is encouraged.
- **EPA has the authority to issue guidelines and community fact sheets for consideration by states and communities in making land use decisions that the Agency believes will reduce pollution.**[50] For example, EPA has used such authority in its Smart Growth recommendations. In certain circumstances, EPA has the authority to address issues associated with the siting of goods movement facilities.

Recommendations. Based on the principles described above, this section identifies specific recommendations for EPA action to directly effect or encourage the needed changes.

22. **EPA should ensure that its staff is familiar with, conversant about, and engaged on local and regional goods movement issues.** Specific steps should include conducting site visits of selected goods movement environmental justice communities to view land uses where significant emissions sources are located near sensitive receptors, so that EPA is as familiar with the goods movement issue as it is with TRI emitters and Superfund sites. By meeting with community leaders and residents, as well as with state and local air pollution regulators, goods movement industry representatives and authorities (port, rail and trucking industry, and distribution center developers), and scientific experts on the health impacts of air pollution, EPA will have a solid basis for moving forward on several fronts. Also, EPA will have established a basis that includes guidance for addressing the relationship between land use and air quality to protect public health and inform future land use with consideration of cumulative impacts.
23. <u>**EPA should develop national guidance for addressing land use decisions and air quality with regard to separating sensitive receptors from mobile source air pollution generated by goods movement facilities, including highways, ports, rail yards, and distribution/transload centers.**</u> For this guidance, EPA could use as background the work done by the HEI reviewing research findings and guidance on suggested buffers developed by the California Air Resources Board (CARB), while recognizing that each goods movement facility has different operational dynamics, and the location and population density of nearby residents can vary widely. EPA already has a document[51] on land use activities and air quality, but it does not mention goods movement nor the

---

[50] See for example, http://www.epa.gov/livability/pdf/whtissg4v2.pdf
[51] See http://www.epa.gov/oms/stateresources/policy/transp/landuse/r01001.pdf, published January 2001

scientific studies about health effects in close proximity to traffic-related pollution, so it needs updating. As a reference, CARB has recommended not siting "new sensitive land uses such as homes, schools, daycare centers, playgrounds or medical facilities near goods movement facilities." Its recommendations included avoiding the "siting of sensitive land uses within 500 feet of a freeway; 1000 feet of a distribution center or a service/maintenance rail yard; and immediately downwind of ports in the most heavily impacted areas." In addition, for "facilities within one mile of a rail yard," CARB recommended that consideration be given to "possible siting limitations and mitigation approaches.[52]" This guidance should include some consideration of site-specific factors and be widely disseminated to the transportation and logistics industry, planning officials, school administrators and boards, real estate developers and others.

24. EPA should develop and publicize a "best practices" clearinghouse, describing successful methods of reducing diesel emissions in each goods movement sector as well as successful methods of engaging communities in that process, including copies of NEPA letters that EPA has developed on goods movement issues. With such information readily accessible, community residents, industry, port and transportation officials will not have to "start from scratch" in researching successful mitigation measures and alternative technologies that they might want to consider when considering land uses.

25. EPA should make publicly available staff comments on NEPA environmental reviews for port, rail or highway facilities publicly available and part of the Goods Movement Clearinghouse as referenced in recommendation 24. EPA should post such comments on each Region's website, with a link to these comments from the Region's EJ page. In this regard, EPA should also consider whether a review or possible update of EPA's 11-year old *Final Guidance for Incorporating Environmental Justice Concerns in EPA's NEPA Compliance Analyses* is needed to address concerns about environmental justice from mobile source air pollution at goods movement facilities.

26. EPA should continue to work with the DOT to update its FHWA guidance to state DOT agencies about methods for quantitatively analyzing mobile source air toxics (MSAT) for new/expanding transportation infrastructure projects, as well as with other DOT agencies (FRA, FAA) for similar guidance on new/expanding rail facilities and airports), including the need to consider the body of data showing health effects in close proximity to traffic-related pollution. This strategy should include developing educational materials on other health-related topics to help the public understand how transportation and land use decisions relate to near roadway health impacts, quality of life issues, and physical activity limitations. Providing this information will make the public better equipped to provide meaningful input during the public participation process.

27. EPA should conduct an analysis of its legal authorities to influence land use decisions on the siting of new or expanded goods movement activities and facilities, including highways..

3.5 TECHNOLOGY

Principles and Framework

- Currently available emission reduction technologies can provide immediate air quality benefits at goods movement facilities. These technologies include energy conversion technologies, fuels, and after-treatment devices.
- Regulatory measures mandating cleaner technologies as new equipment enters mobile fleets do not support the pace of change that impacted communities expect for cleaner air. Non-traditional technological approaches can further reduce goods movement related emissions. These approaches include use of emissions capture technologies, renewable energy sources, expanded electrification, and hybridization.
- Technologies are available to improve goods movement facility air quality extend beyond mobile equipment to infrastructure and systems that improve facility efficiency and throughput. These approaches include the use of radio-frequency identification devices and GPS-based automation and optimization of product movement, automated vehicle processing, and other systems that improve

---

[52] California Air Resources Board. *Air Quality and Land Use Handbook: A Community Health Perspective.* May 2005. Available at: http://www.arb.ca.gov/ch/handbook.pdf

overall system efficiency. Such technology can provide a robust data source to achieve the goal of transparency and meaningful public involvement
- EPA attention to environmental justice goods movement issues provides an opportunity for the Agency to support development of world-leading technological innovation that can provide further emission improvement opportunities.

Please see EPA's Clean Air Act Advisory Committee Clean Diesel Report,[53] EPA's Ports Strategy[54], the California Air Resources Board Draft Report – *Technical Options to Achieve Additional Emission and Risk Reductions from California Locomotives and Rail yards*,[55] various industry documents, and other references that detail the wide range of technologies available to improve air quality in and around goods movement facilities.

Recommendations. Based on the principles described above, this section identifies specific recommendations for EPA action to directly effect, or influence, the needed changes.

28. <u>EPA should expand the amount of credit allowed in SIPs that drive states to offer economic and other incentives to reduce existing equipment emissions through accelerated deployment of cleaner technologies.</u> Such programs must include enforceable provisions that provide certainty to impacted communities that those emissions benefits will be achieved. This guidance should encourage the development of programs which offer sufficient incentives that encourage equipment owners to pick up a substantial portion of costs in order to extend the life of an existing piece of equipment with lower emitting technologies. This guidance also should encourage the adoption of technologies and methodologies expediting vehicle, container, and other product movement through goods movement facilities.
29. <u>EPA should establish, within a national clearinghouse, information about goods movement emissions reduction technologies, techniques, and best practices.</u> EPA's guidance development for best practice mitigations should be incorporated into all new goods movement facility and corridor projects. These practices should help land use planners, infrastructure developers, and others identify the cleanest available technologies appropriate to the specific nature of a given goods movement development. EPA should make such a clearinghouse available to affected communities to inform and empower local communities to address projects under review for mitigation.
30. <u>EPA should use its own research and development resources, as well as partner with other federal partners and other stakeholders, to develop and accelerate the commercialization of innovative technologies that will benefit communities impacted by goods movement activities.</u>

3.6 ENVIRONMENTAL PERFORMANCE, PLANNING, AND MANAGEMENT

Environmental management and planning tools relevant to measuring and reducing environmental justice impacts of goods movement include environmental management systems (EMS), clean air action plans, emissions inventories, facility air monitoring, emissions reduction agreements such as SmartWay, and performance standards for operations.

EPA has various programs underway which are focused on improving the environmental performance of public and private organizations involved in goods movement through non-regulatory initiatives. Some of these are specific to the goods movement industry, such as the SmartWay Transport Partnership, while others are more general but include affected activities in the goods movement sector, such as the Clean Diesel Campaign and the Sector Strategies Program activities with ports.

---

[53] See http://www.epa.gov/air/caaac/pdfs/2007_01_diesel_rec.pdf
[54] See http://www.epa.gov/ispd/ports/#ports
[55] See http://www.arb.ca.gov/railyard/ted/122208ted.pdf

EPA has taken a leadership role in its Sector Strategies program to encourage development of EMS in certain aspects of the goods movement system, particularly ports, through the development of training materials and assistance in funding of training programs. In certain goods movement industry sectors, EPA has encouraged the adoption of environmental management tools that address specific environmental aspects of goods movement without adopting a full EMS. For example, EPA's SmartWay program has taken a leadership role in providing a Freight Logistics Environmental and Energy Tracking (FLEET) Model to shippers, truck operators, and rail carriers to assess their corporate emissions footprints. Additionally, the organizations use the FLEET model to project the amounts of reductions that are possible with different technology and implementation options; allowing operators to customize their reduction strategy.

Principles and Framework. Goods movement activities in any particular location involve many public and private organizations, while most existing environmental planning and management systems (i.e., EMS, and voluntary industry partnerships) only affect individual organizations or sectors. The holistic assessment and coordinated reduction of environmental justice impacts from the larger goods movement sector is rare – due to the involvement of many public and private organizations in any given location and the lack of an obvious "home" or regulatory driver. Perhaps the best examples of holistic plans are the clean air plans that are being applied in certain port areas (i.e., the Los Angeles-Long Beach and the Seattle/Tacoma/ Vancouver BC ports); however, even these ambitious plans are limited to goods movement impacts directly associated with ports and not the ancillary infrastructure supporting the ports. Therefore, the plans are not comprehensive.

Key principles guiding action in this area are:

- **EMSs are an established way of improving environmental performance beyond regulatory compliance.** Numerous public and private organizations have found that EMS provides a structure that makes business sense as well as reduces environmental impacts. EMSs require senior management support and involvement, including approval of the environmental aspects and impacts of the organization and periodic review of progress and results (typically at least twice per year).
- **Environmental justice issues can be readily incorporated in EMS planning efforts as an aspect of an organization's activities.** Identifying environmental justice as an "aspect" of an organization's activities in an EMS would require the organization to go through the process, develop required plans, establish specific objectives and targets, implement the plan, and monitor and track the results then set new targets to further minimize or eliminate the impact (i.e. continuous improvement). In other cases companies or institutions may have departments dedicated to community outreach and response, or make other allocations of responsibility within the organization, and those staff would be best positioned to develop goals to address environmental justice—and should be encouraged to do so. These staff should be encouraged to look to EMS principles in terms of methodical review, goal setting, and tracking.
- **Management tools for improving environmental performance have not been applied consistently in goods movement.** Rather, they have been applied to greater and lesser degrees in certain industry sectors.
- **EMS use in the private sector is a useful tool to improve environmental performance but is generally used only internally due to integration of business confidential information.** This limits the ability to integrate EMS between public and private sector participants in goods movement.

> **Environmental Management Systems (EMS)**
> An EMS is a management system that allows an organization to systematically manage its environmental impacts by incorporating environmental considerations and decision-making into an organization's daily operations and long-term planning. An EMS is a continual cycle of planning, implementing, reviewing, and improving the processes and actions that an organization undertakes to meet its business and environmental goals.

Recommendations. Based on the principles described above, this section identifies specific recommendations for EPA action to directly effect, or influence, the needed changes.

31. <u>EPA should, through its SmartWay and other programs, encourage shippers, trucking firms, and railroad companies to use corporate modeling and management tools like the FLEET model and EMSs to measure their environmental footprints.</u> EPA should continue to develop additional tools and models and encourage the use of EMSs for other segments of the goods movement system, including ocean-going carriers, air carriers, major developers of distribution centers, state transportation departments, and municipal planning organizations. EPA's involvement in training can help encourage both the development of EMS for general environmental improvement as well as specific guidance on including environmental justice concerns in the EMS planning process. Through the trainings, EPA should encourage public participation in public entity EMS planning (both initially and as part of the periodic review process where results are publicly reported and the plan modified as needed) and encourage integration of relevant portions of private sector EMS or other tools where the private sector entities are willing to do so.
32. <u>EPA should provide technical assistance funding to review environmental management practices of organizations involved in goods movement in geographic areas with environmental justice concerns.</u> Coordinated reviews could help identify potential synergies or conflicts between various management approaches, which could serve as part of the "check" process of continuous environmental improvement.
33. <u>EPA should develop and provide educational material, programs, and funding to organizations which could help develop a more comprehensive approach to emission reductions due to their areas of authority.</u> In particular, municipal and regional planning organizations and transportation departments have relevant responsibilities but may lack training and awareness of environmental justice impacts of goods movement facilities. This effort should include both information targeted at senior management and elected officials as well as expansion of the technical guidance that EPA has developed relevant to assessment and reduction of environmental justice impacts of certain goods movement industry sectors so that it is relevant to more goods movement industries and participants.
34. <u>EPA should encourage the funding of pilot projects, which utilize a holistic approach and the reduction of environmental justice impacts from goods movement in specific geographic areas.</u> EPA's involvement in this effort should also encourage public participation in EMS planning (both initially and as part of the periodic review process where results are publicly reported and the plan modified as needed). EPA should allow funding of these kinds of holistic environmental justice impact reduction plans for goods movement as Supplemental Environmental Projects for settlement of enforcement actions. Where EPA funding is not available, EPA should encourage other Federal, State, and local governments as well as private entities to fund such projects.

3.7. RESOURCES, INCENTIVES, AND FINANCING

Principles and Framework. Funding and financing tools exist at the federal, state, and local levels that target solutions to improve air quality in environmental justice communities. Solutions cannot be solely provided by government resources. Timely and comprehensive solutions must include both government and private resources. However, the existing funding and financing tools have not been fully used or lack the ability to leverage private resources to alleviate air quality problems in these communities. With budget constraints at all levels of government, focus should be placed on directing existing funding and financing tools toward communities with high pollution levels and environmental justice issues. Existing tools include a variety of cleanup programs at EPA Community Development Block Grants and a variety of state and federal tax credit programs, including the New Markets Tax Credit Program administered by the U.S. Treasury Department's Community Development Financial Institutions Fund (CDFI Fund) and targeted toward business and projects located in low-income communities

Key principles guiding action in this area are:

- **Current resources available to mitigate the impacts of diesel emissions from goods movement are insufficient to ensure environmental justice.**
- **There are insufficient mandatory allocations of Federal transportation and infrastructure funds for cost-effective air quality projects.** Increases to such allocations could be used to address mitigation options that would improve air quality in the community.

- **Proposed projects, including goods movement expansions, do not accurately account for environmental impact-related costs.** Internalization of all environmental mitigation costs, as well as other project costs, must be included in the final project budget.
- **New sources of funding and new financing programs are needed to mitigate goods movement air emissions in environmental justice communities.** Funding and financing tools could include: fees (local, state, or Federal), surcharges, tax credits, tax-exempt bonds, and loan guarantees.
- **Supplemental resources needed to reduce emissions to a desired level, beyond those provided by Federal and state programs, should be financed by regional and local tools that are agreed upon by all stakeholders in an open collaborative process, such as the community facilitated strategy and collaborative governance approaches described in 3.1.** Several existing financing tools could successfully be applied to address air quality issues in environmental justice communities.
- **Incentives to encourage actions by private entities involved in goods movement that go beyond regulatory minimums should be provided to assist in meeting additional costs for reducing exposure and risk to impacted communities.**

Recommendations. Based on the principles described above, the following identifies specific recommendations for EPA action to directly effect, or influence, the needed changes.

35. <u>EPA, in partnership with other federal agencies, should propose increased funding for programs that encourage the accelerated development and deployment of lower emitting technologies and effective mitigation strategies into the goods movement sector.</u> EPA should prioritize use of National Clean Diesel Campaign funding to improve the air quality within goods movement impacted communities by promoting the deployment of cleaner technology using certified and verified technologies. EPA should provide factual information about the national cost to modernize the entire goods movement fleet, the health and economic benefits of accelerating that modernization, and the possible mechanisms to help incentivize that effort
36. **EPA should seek full funding for the *Diesel Emission Reduction Act* of 2005[56] at the full authorized level, with monies directed to areas with high health impacts from goods movement activities.** EPA, in its prioritization of grant awards, should ensure that these funds and the allocation formula used for these funds is based on reducing risk in environmental justice communities impacted by goods movement activities. EPA should work with Congress, DOT, and other federal agencies related to goods movement activities, to ensure that any new fees considered for cargo or freight infrastructure include funding to reduce emissions and health risk.
37. <u>EPA should seek joint innovative financing strategies with other Federal agencies, non-profit organizations, and private industries.</u> These financing strategies should encourage public-private partnerships that provide flexible financing options as well as informational outreach and technical assistance. Key stakeholders to include in such partnerships are: other Federal agencies; state and local governments/agencies; business and finance partners, including non-profit lenders; and community environmental justice and other organizations
38. <u>EPA should seek to create incentives for facilities and participants in potential public-private partnerships</u>. Incentives should be both financial- and compliance-based and include community involvement in determining where funds are to be used for mitigation in these communities. Banks should be encouraged to provide loans that target and alleviate the negative impact of goods movement. Banks should receive Community Reinvestment Act credit for the transactions.
39. <u>EPA, in partnership with other federal agencies, should encourage the funding of projects to clean up the legacy diesel fleet and mitigate impacts on communities.</u> Such incentives include but are not limited to:
    - Publicize emissions mitigation from goods movement as a qualifying Supplemental Environmental Project (SEP) if proposed by regulated sources to settle environmental violations near environmental justice communities;

---

[56] See Energy Policy Act of 2005 (Pub.L. 109-58)

- Leverage DOT Congestion Mitigation and Air Quality funding for cost-effective air quality projects that directly reduce emissions from diesel vehicles and equipment, and push for set asides from other Federal funding for infrastructure;

40. **EPA, having already endorsed the recommendations of the Environmental Financial Advisory Board (EFAB) report to establish State Air Quality Finance Authorities that would assist owners of small fleets of diesels and of small goods movement related businesses to receive low cost financing, should work with States and Congress to implement these recommendations.**
    - EPA and DOT should agree to set aside a significant portion of DOTs allocation of Private Activity Bond authority for projects related to goods movement emissions mitigation.

41. **EPA should support access to financing programs (such as loans or loan guarantees) for entities that may have to comply with future federal or state emissions regulations.**

APPENDICES

THIS PAGE INTENTIONALLY LEFT BLANK

A Federal Advisory Committee to the U.S. Environmental Protection Agency

APPENDIX A

NATIONAL ENVIRONMENTAL JUSTICE ADVISORY COUNCIL

CHARGE FOR DEVELOPING RECOMMENDATIONS TO ADDRESS THE AIR QUALITY IMPACTS OF GOODS MOVEMENT ON COMMUNITIES

ISSUE

Environmental pollution from the movement of freight is becoming a major public health concern at the national, regional and community levels. Also known as "goods movement," the distribution of freight involves an entire system of transportation facilities, including seaports, airports, railways, truck lanes, logistics centers, and border crossings. The distribution of goods involves diesel-powered vehicles and equipment almost every step of the way, resulting in significant emissions of particulate matter (PM), nitrogen oxides (NOx), hydrocarbons, and other air toxics throughout the process. A substantial body of scientific evidence asserts these emissions are or could be linked to respiratory disorders, cancer, heart disease, and premature death. EPA's *Health Assessment Document for Diesel Engines* (EPA, May 2002) and its *Regulatory Impact Analysis for Heavy Duty Engine and Vehicle Standards and Highway Diesel Fuel Sulfur Control Requirements* (EPA December, 2000) define agency public health concerns surrounding existing diesel engine emissions. In addition, community concerns include traffic congestion, noise, pedestrian safety, and overall community aesthetics and land use considerations.

The environmental, public health, and quality-of-life impacts of goods movement on communities are more pronounced in areas with major transportation hubs and heavily trafficked roads. "Near roadway hot-spots" – localized areas with elevated levels of air pollution – is an issue of long-standing concern to EPA and other environmental health agencies. This issue also is a matter of increasing concern to government transportation and planning agencies. Research shows that the many communities, including minority and/or low-income communities, living near these transportation hubs and thoroughfares, already bear disproportionate environmental impacts because of their close proximity to multiple pollution sources.

Recent and projected increases in foreign trade require significant improvements to the essential infrastructure needed to move freight from coastal ports to the rest of the country. For example, the American Association of Port Authorities estimates that the amount of cargo handled by American seaports, currently about 2 billion tons a year, will double in the next 15 years. In most cases, seaports are just the first stop. It has been argued that if the continued investment in goods movement infrastructure does not simultaneously address the serious environmental and/or public health concerns associated with goods movement, the already high levels of air pollution and their associated health effects will increase and further harm public health and quality-of-life. It is becoming increasingly important that these entities operate sustainably, i.e., economically viable, environmentally and socially responsible, safe, and secure.

In accordance with Administrator Johnson's memorandum, "Reaffirming the U.S. Environmental Protection Agency's Commitment to Environmental Justice" (November 4, 2005), EPA maintains an ongoing commitment to ensure environmental justice for all people, regardless of race, color, national origin, or income. In years past, EPA has made substantial efforts to address environmental justice concerns related to air pollution issues. The National Clean Diesel Campaign, utilizing strategies such as diesel retrofits and anti-idling technologies, and the Community Action for a Renewed Environment (CARE) initiative, are but two programs that EPA's Office of Air and Radiation (OAR) has developed to respond to the environmental justice issues associated with air pollution concerns. EPA has strategically focused its clean diesel efforts on five key sectors: school buses, ports, construction, freight, and agriculture. These sectors represent the diverse array of diesel engines in use today and provide the best

---

A Federal Advisory Committee to the U.S. Environmental Protection Agency

opportunities for EPA to obtain emissions reductions from existing engines that can significantly protect public health. EPA also has developed several innovative financial models that have the potential to upgrade many of the trucks and other diesel equipment that move our nation's goods, if low cost financing can be obtained. Other OAR programs like the SmartWay Transport Partnerships and Agency programs like OPEI's Sector Strategies Program for Ports, also have contributed to addressing the environmental health impacts of goods movement on communities, including minority and/or low-income communities.

As an important first step, EPA also has been encouraging ports to do emission inventories, as this provides a baseline from which to create and implement emission mitigation strategies and track performance over time. This can be accomplished within the framework of a company's Environmental Management System (EMS), which also fosters a company culture of environmental stewardship. In addition, EPA is addressing emissions from new engines. The new standards for highway diesel engines are expected to reduce the emissions of individual diesel vehicles dramatically, with stringent PM and NOx emission standards beginning in 2007 and 2010 model years, respectively. Stringent non-road diesel engine standards phase in between 2008 and 2014. On March 2, 2007, the Administrator also proposed more stringent standards to reduce the PM and NOx emissions of locomotive and marine diesel engines.

Administrator Johnson's November 2005 memorandum also directed EPA offices to: (1) establish, as appropriate, measurable environmental justice commitments for eight national environmental justice priorities; and (2) identify the means and strategies to achieve the commitments and measure outcomes to help ensure that Agency resources reach disproportionately burdened communities, including minority and/or low-income communities. EPA's national environmental justice priorities relevant to this charge include: Reduce Asthma Attacks; Reduce Exposure to Air Toxics; and Collaborative Problem-Solving. Additionally, two priorities in Administrator Johnson's Action Plan pertain to diesel emissions reduction and SmartWay Transport.

THE CHARGE

EPA requests that the NEJAC provide advice and recommendations about how the Agency can most effectively promote strategies, in partnership with federal, state, tribal, and local government agencies, and other stakeholders, to identify, mitigate, and/or prevent the disproportionate burden on communities of air pollution resulting from goods movement.

As it considers this question, the NEJAC may wish to undertake the following activities or approaches:

- Through literature review and community input, identify and summarize the most significant community environmental and/or public health concerns related to air pollution from goods movement activities.

- Identify and summarize the types of data and tools that can be used to determine the location and magnitude of disproportionate impacts of air pollution related to goods movement activities on communities, and recommend ways in which the Agency can promote more effective utilization of such data and tools.

- Identify the key lessons learned regarding strategic alignment, collaboration, and partnerships to mitigate and/or prevent environmental and/or public health impacts on communities that could be replicated in areas affected by air pollution related to goods movement. Specifically, identify the venues and other mechanisms that EPA can use to work with other government agencies, industry, and communities, in areas such as environment, public health, and/or transportation, to reduce community exposure to air pollution from goods movement activities.

- Develop and recommend strategies for EPA and partners which utilize and promote meaningful community involvement in federal, state, tribal, and local government decision-making processes to address local environmental health impacts of goods movement. Specifically, identify strategies in such areas as environment, public health, and/or transportation, and those procedures for proposing and building new infrastructure related to goods movement. Agencies may include port authorities, federal and state departments of transportation, the U.S. Army Corps of Engineers, and metropolitan planning organizations.

- Develop a tool box of strategies that EPA and its government, industry, and community partners can promote to enhance current approaches (e.g., anti-idling, buy-outs of old trucks, capital investments to provide cleaner trucks, diesel collaboratives, and the CARE and Congestion Mitigation and Air Quality Improvement programs) being pursued which address community concerns related to goods movement. Such strategies could include the identification of existing, and the creation of, new community development financing programs that provide low-cost financing to businesses operating in environmentally sensitive areas. An example of a facility-based strategy is the development of an EMS that can be utilized to assess, address, and measure progress about air quality or other environmental and human health issues.

NEJAC Charge to the Goods Movement Work Group

Draft a report for consideration by the NEJAC Executive Council to document the significant impacts of air pollution resulting from goods movement activities and their incremental increase with projected growth. The draft report may include suggestions about how EPA can most effectively promote strategies, in partnership with federal, state, tribal, and local government agencies, and other stakeholders, to identify, mitigate, and/or prevent the disproportionate burden on communities by air pollution resulting from goods movement. The draft report should reflect the perspectives of all stakeholder groups and should reflect an effort to answer the following questions:

(1) What are the most significant community environmental and/or public health concerns related to air pollution from goods movement activities;
(2) How can information resources be used to better identify and assess the population segments or communities that are likely to bear the maximum burden of impacts;
(3) What strategies can EPA pursue to ensure mitigations of impacts and promote collaborative problem-solving and meaningful community involvement in the decision-making processes at the federal, state, tribal, and local government levels; and
(4) What strategies can stakeholders pursue to ensure emissions reductions, including but not limited to financing options, technological solutions, land use guidelines, as well as regulatory mechanisms.

THIS PAGE INTENTIONALLY LEFT BLANK

A Federal Advisory Committee to the U.S. Environmental Protection Agency

APPENDIX B

GLOSSARY of ACRONYMS and TERMS

Air Toxics – Pollutants that are known or suspected to cause cancer or other serious health effects (also known as hazardous air pollutants)

CAAAC – Clean Air Act Advisory Committee, a federal advisory committee of the U.S. Environmental Protection agency. The CAAAC provides independent advice and counsel on the development of policy and programs necessary to implement and enforce the requirements of Clean Air Act amendments enacted in 1990. The Committee is consulted about economic, environmental, technical, scientific, and enforcement policy issues.

CARB – California Air Resources Board, state air regulatory agency that is a part of the California Environmental Protection Agency, an organization which reports directly to the Governor's Office in the Executive Branch of California State Government. The mission of the Board is to promote and protect public health, welfare, and ecological resources through the effective and efficient reduction of air pollutants while recognizing and considering the effects on the economy of the state.[57]

CASAC – Clean Air Act Scientific Advisory Committee, a federal advisory committee of the U.S. Environmental Protection Agency. The CASAC provides independent advice on the scientific and technical aspects of issues related to the criteria for air quality standards, research related to air quality, source of air pollution, and the strategies to attain and maintain air quality standards and to prevent significant deterioration of air quality.

Conformity – Transportation conformity requires that new projects relying on Federal funding or approval are consistent with air quality goals. General conformity applies to projects to site, modify, or expand federal facilities (like military bases) and facilities relying on Federal funding or approval, like seaports and airports. Under the Federal Clean Air Act, both types of conformity are designed to ensure that these activities do not worsen air quality or interfere with the attainment of the National Ambient Air Quality Standards.

Drayage – Drayage typically refers to hauling containers or other cargo by truck at marine terminals or intermodal facilities.

EMS – Environmental Management System, a set of processes and practices that enable an organization to reduce its environmental impacts and increase its operating efficiency.

Gross Emitter – Vehicles that violate current emissions standards applicable to that vehicle and that have emissions that substantially exceed those standards.[58]

Goods Movement – Goods movement refers to the distribution of freight (including raw materials, parts, and finished consumer products) by all modes of transportation, including marine, air, rail, and truck.

PM – Particulate Matter

$PM_{2.5}$ or PM2.5 – Particulate matter equal to or smaller than 2.5 micrometers, also known as fine particulate matter

$PM_{0.1}$ or PM 10 – Particulate matter equal to or smaller than 100 nanometers, also known as ultra-fine particulate matter

---

[57] http://www.arb.ca.gov/html/mission.htm
[58] http://www.epa.gov/ems/

PPM – parts per million

NOx – nitrogen oxide, a generic term for mono-nitrogen oxides produced during combustion, especially combustion at high temperatures.

SOx – sulfur oxide, a generic term describing emissions to air that mainly come from the combustion of fossil fuels containing variable proportions of sulfur.

µg/m^3 – Micrograms per cubic meter

VOC – Volatile Organic Compounds

WHO – World Health Organization, which is the directing and coordinating authority for health within the United Nations system. It is responsible for providing leadership on global health matters, shaping the health research agenda, setting norms and standards, articulating evidence-based policy options, providing technical support to countries and monitoring and assessing health trends.[59]

Yard equipment – Mobile cargo handling equipment is any motorized vehicle used to handle cargo delivered by ship, train, or truck, including yard trucks, top handlers, side handlers, reach stackers, forklifts, rubber-tired gantry cranes, dozers, excavators, loaders, mobile cranes, railcar movers, and sweepers[60].

---

[59] http://www.who.int/about/en/
[60] California Air Resources Board. Regulation for Mobile Cargo Handling Equipment at Ports and Intermodal Rail Yards. February 2007. Available at: www.arb.ca.gov/ports/cargo/documents/chefactsheet0207.pdf

APPENDIX C

Summary of EPA FACA Recommendations Regarding Air Quality From Freight Movement and Environmental Justice

CHILDREN'S HEALTH PROTECTION ADVISORY COMMITTEE	CLEAN AIR ACT ADVISORY COMMITTEE	GOOD NEIGHBOR ENVIRONMENTAL BOARD
Children's Health and Climate Change – 8/30/05 Letter - EPA should use existing regulatory authority to require mandatory controls on US GHG emissions. - EPA needs to explore regulatory and other actions, alone or in partnership with the Department of Transportation, to reduce GHG emissions from mobile sources. - EPA should expand development of new technology through the Climate Change Technology Program, economic incentives, and other means, including technologies in energy generation, domestic and commercial energy use, mobile sources, manufacturing, and waste generation and management. **Lead Standard – 2/2/07 Letter** Not to revoke the current lead standard in NAAQS **Review of the NAAQS for Ozone Letter 3/23/07** - We urge that the lower- and more child protective- value of 0.060 ppm be selected from the range suggested by the CASAC. - We support the form of the new standard to be specified to the thousandths of ppm. - Children experience a wide variety of health impacts from ozone exposure that should be recognized in considering benefits from lowering the 8 hour ozone standard.	**Air Quality Management Subcommittee Recommendations to the CAAAC: Phase II, June 2007** - EPA, state, local governments, and tribes should adopt a comprehensive AQM planning process, and, through this process, create plans to move from a single pollutant approach to an integrated, multiple pollutant approach to managing air quality. - Improve environmental and health data - Take climate change into account - Support transportation and land use scenario planning - Integrate air quality planning into land use, transportation, and community development plans - Provide incentives for voluntary and innovative land use and transportation approaches **National Clean Diesel Work Group Recommendations for Reducing Emissions from the Legacy Diesel Fleet Presentation – 1/11/07** - Port Sector - Develop Emissions inventory - Port Sector - Share Best Practices: Educational Materials and Tools - Freight Sector - Increase demand for cleaner, more efficient freight	**9th Report: Air Quality and Transportation & Cultural and Natural Resources on the U.S.-Mexico Border** - Border Stations and Transportation Infrastructure: Bolster infrastructure, technology, personnel and related activities through substantial new funding, and intensify long-range planning and coordination at the bi-national, national, state and locals levels to cope with the congestion at border crossings, and thus reduce air pollution. - Emissions: Harness new and emerging technologies and fuels to reduce emissions from diesel trucks, buses, municipal and private fleets and passenger vehicles, and identify private/public funding sources to accelerate the process. - Public Transit and Alternatives to Driving Alone: Encourage public transportation, ridesharing, car-sharing, biking, and walking in border cities so that fewer people will drive alone, thus reducing motor vehicle trips and the emissions of pollutants.

ENVIRONMENTAL FINANCE ADVISORY BOARD	GOVERNMENTAL ADVISORY COMMITTEE
Letter to Administrator Johnson, dated 11/1/07, and Report on Innovative Financing Programs for Air Pollution Reductions - EPA should develop a revolving loan fund for air quality modeled after the Clean Water State Revolving Loan and Safe Drinking Water State Revolving Fund. - EPA should encourage states to create Air Quality Finance Authorities (AQFA) or empower existing environmental finance authorities to finance certain types of air emission reduction equipment; or, at least create a state-wide or regional air emission reduction financing program. These AQFAs would issue lower interest bonds for emission reduction equipment as well as offer discounted price on SmartWay kits or other similar products. - EPA approach DOT regarding the use of the untapped $15 billion in private activity bonds to underwrite mobile source air emissions reduction efforts if this can be done on terms consistent with title 23 of the US Code.	**Letter to Administrator dated 05/10/06** - Renewable Energy Markets and Clean Fuels: The GAC supports and encourages more EPA and CEC projects that promote the use of cleaner fuels and development of renewable energy markets in North America. - Increase CEC Emphasis on Renewable Energy: The committee supports the work of the CEC on renewable energy and recommends an increasing emphasis on projects that would reduce reliance on fossil fuels and increase the availability and use of renewable energy sources. - Climate Change Impacts on Indigenous Peoples: Climate change has caused, and may cause further impacts to the traditional lifestyles of North America's indigenous peoples regardless of whether they live in the far northern hemisphere or hot, dry desert regions. The GAC expresses its profound concern about such social impacts to indigenous peoples in North America and encourages the U.S. Government to be acutely aware of such implications and consider the consequences of climate variability on Native Americans

www.ingramcontent.com/pod-product-compliance
Lightning Source LLC
Chambersburg PA
CBHW081625170526
45166CB00009B/3105